ECOLOGISTS AND
ENVIRONMENTAL POLITICS

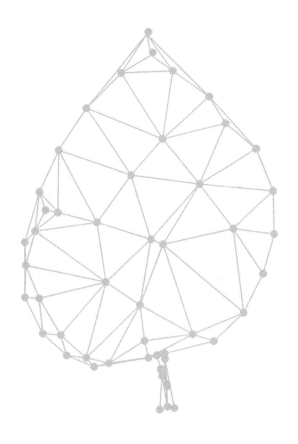

ECOLOGISTS AND ENVIRONMENTAL POLITICS

A HISTORY OF CONTEMPORARY ECOLOGY

STEPHEN BOCKING

WEST VIRGINIA UNIVERSITY PRESS • MORGANTOWN 2017

To Rita, Paul, and Emma

First edition copyright © 1997 by Yale
University. Second edition copyright © 2017
by West Virginia University Press.

ISBN:
PB: 978-1-943665-64-8
EPUB: 978-1-943665-65-5
PDF: 978-1-943665-66-2

Library of Congress
Cataloging-in-Publication Data for the first
edition

Bocking, Stephen, 1959–
Ecologists and Environmental Politics: a
history of contemporary ecology / Stephen
Bocking.
p. cm.

Includes bibliographical references (p.) and
index.
1. Ecology—History. 2. Environmental
policy—History.
I. Title.
QH540.B63   1997
574.5'09—dc20   96–34071
                              CIP

Contents

Preface to the 2017 Edition   vii

Preface   xi

Abbreviations   xv

1 Introduction: Ecologists, Institutions, and Society   1

Part One / Great Britain

2 Origins of the Nature Conservancy   13

3 Woodland Ecosystems and Nature Reserves   38

Part Two / United States

4 Ecosystems and the Atom at Oak Ridge National   63
   Laboratory

5 Oak Ridge Ecosystem Research and Impact   89
Assessment

6 Forest Experiment and Practice at Hubbard Brook   116

Part Three / Canada

7 Ecology and the Ontario Fisheries   151

8 Comparing Ecologists and Their Institutions   179

9 Understanding Ecology's Place in the World   203

   Notes   207

   References   236

   Index   266

Preface to the 2017 Edition

Like ecology, this book was a product of its times. In the late 1980s I was a grad student in history of science at the University of Toronto, recently arrived from biology (specifically, ecology and limnology). I studied first ideas about evolution after Darwin, such as those of Alpheus Spring Packard, an American who interpreted cave fauna that lacked eyes and other adaptations to light as evidence of the inheritance of acquired characteristics. But I soon shifted to the history of ecology, intrigued by the interesting fact that the discipline took early root in the Great Lakes region. I pursued this by comparing the work of Stephen Forbes (later famous for his 1887 essay "The Lake as a Microcosm") in Illinois and Jacob Reighard of the University of Michigan. By the time I had begun my PhD, my interests had shifted forward to more recent debates about ecology and its social roles. So I turned to the development of ecology since the Second World War. That work became my dissertation, which, after more years of writing, became *Ecologists and Environmental Politics.*

At that time few historians had written about ecology. Frank Egerton had produced several useful studies and surveys, Sharon Kingsland had examined population ecology (and had just graduated from Toronto a few years before), Robert McIntosh had produced a book-length survey of the "background" of ecology, and, most ambitiously, Donald Worster had written a history of ecological ideas. But I was surprised that hardly anyone (apart from Worster) had paid attention to what I considered one of the central features of ecology: its relation to environmental concerns. We were once supposed to have entered an "Age of Ecology"; I had also read descriptions of ecology as the "intellectual foundation" of the environmental movement—and even as a "subversive science." So I wondered about that. Had ecology become not just a scientific discipline, but a source of guidance—even a conscience—for society? And if it had, how had that affected ecology itself?

Forming these questions was easy; working out how to answer them took more time. There were few models. Worster had presented a synthetic perspective on ecology and environmental values, contrasting Arcadian coexistence and Imperial domination. Historians of environmental thought viewed ecology mainly as a source of values, with ecologists' social roles interpreted in terms of a few key ideas such as stability, competition, chaos, or cooperation. Policies that invoked ecology, such as ecological modernization, similarly presented it as a source of essential principles, not a living scientific discipline.

So these perspectives didn't speak to my interest in ecology as something done by real people. I was curious about ecologists' working life, partly because this was what I'd most enjoyed as a student (especially messing around with boats and sampling equipment). It also seemed to me that if social considerations influenced ecology, and vice versa, then this should be evident in how ecologists did the practical work of assembling, testing, and applying their knowledge. It should also be evident in where ecologists worked: their results and relevance seemed to have a lot to do with where they located their research—in forests, lakes, a desert, or elsewhere. I'd also noticed that other disciplines, such as toxicology or chemistry, often seemed more relevant than ecology to environmental concerns—suggesting that there was no necessary link between ecology and environmental affairs.

These notions also reflected the state of the history of science in the 1980s. Interest was shifting from conceptual developments to practices: that is, how, specifically, scientists construct knowledge in the lab, field, and elsewhere. From Peter Galison's *How Experiments End* I learned about the practice and organization of big science. Martin Rudwick's *The Great Devonian Controversy* and Mary P. Winsor's course at Toronto on "The

Invention of Biology" taught me that disciplines do not simply reflect how nature is organized, but scientists' choices about how to manipulate nature. From Charles Rosenberg's *No Other Gods: On Science and American Social Thought* I took instruction on how scientists' practices and disciplinary ambitions are refracted through institutions. In *Beauty, Health, and Permanence* Samuel Hays traced environmental concerns to peoples' life experiences—a view that complemented a focus on scientists' practices. I also studied political science, science policy, and organizational studies to understand how scientists work in relation to institutions, funding agencies, and government machinery.

Equipped with these perspectives, my next step was to decide where to study. I settled on three places: the Nature Conservancy of Great Britain, the Oak Ridge National Laboratory in Tennessee, and the Hubbard Brook Ecosystem Study in New Hampshire. I'd known about each, and their importance to ecology, since I was a science student. British ecologists had in the 1940s urged creation of the Nature Conservancy, an institution to acquire and manage nature reserves and support their research. Since the 1950s ecological research had developed at Oak Ridge that emphasized ecosystem study and the laboratory's nuclear mission. And at Hubbard Brook ecologists supported by the National Science Foundation studied a watershed ecosystem, drawing implications for forestry and other issues. While all part of postwar Anglo-American ecology, these approaches differed in terms of structure, relations to practical concerns and other disciplines, and—what I found especially interesting—ideas about landscape, with the British view of nature as including humans contrasting with the American wilderness tradition.

At each institution ecologists formed distinctive ties to local places, with field sites and practices linked to scientific goals and environmental concerns. These outdoor "laboratories" became places of experimentation. The Oak Ridge Reservation became a research instrument, with a contaminated lake bed, a forest, and other habitats used to study nutrient dynamics and the transport of radionuclides in ecosystems. Hubbard Brook scientists studied in forests and streams the movement of elements and the consequences of forest cutting and acid rain. British ecologists used field sites to study forest regeneration and ecological restoration, among other topics. Each place also exemplified distinctive political and institutional cultures, including specific environmental concerns, as well as ideas about the appropriate role of government and of experts in managing nuclear technology, forests, and nature conservation. In each case institutions served as gatekeepers between society and ecologists: assembling social factors into particular combinations that influenced the development of ecology (and other disciplines) at each site. The cultures instilled within each institution shaped ecologists' choices regarding their research questions and practices, and their positions in relation to environmental concerns. Even within each institution, these choices could be diverse: in no way were they guided by a single vision. Taking a lesson from Galison's study of big physics, I mapped the divisions of labor at each site: between ecologists focused on just birds, insects, or small mammals and others who studied entire ecosystems; or between those developing practical applications and those preferring to speak mainly to other ecologists.

Working all this out confirmed for me the contribution of historical comparisons to forming a broader understanding while respecting the particularities of place. I had been surprised to find few examples of comparative history of science, although Jonathan Harwood's study of German and American genetics was a useful model. (I'd also already seen the value of comparative analysis when I studied the history of ecology in Illinois and Michigan.) These institutions proved to be ideal for comparison, serving as a natural experiment that conveyed a sense of the diversity of ecologists' experiences during the environmental era. Later, as I revised my thesis for publication, I added another case: ecological and fisheries research at the University of Toronto. Canada presented its own political culture of science and environmental affairs, while fish populations and the Great Lakes presented distinctive environmental challenges, enriching the comparative analysis.

So I found that there had been no single relation between ecology and environmental affairs. Rather, these relations reflected the many local circumstances, demands, and choices made by ecologists and by those who used their results. This corroborated nicely, I thought, William Cronon's comment in his 1993 essay "The Uses of Environmental History": that there is not One Big Problem called "The Environment," but a near infinitude of relations between people and the world. Even amidst all the research accomplished over the last twenty years on the history of ecology, field sciences, conservation biology, and other topics, this observation has stood up rather well.

Stephen Bocking
September 2016

Preface

Many see the science of ecology and environmental politics as inseparable: environmentalism draws its evidence from ecological knowledge, and environmental concerns motivate ecologists in their research. But to ecologists themselves, their place in environmental politics is less straightforward; indeed, they have debated it for decades. It raises for them questions of research priorities, funding, organization, even the definition of their discipline and its social relevance. Many ecologists have found, for example, that even as the environment has become of greater concern to society, it has become more difficult to assert a leading role in addressing this concern. As an essential aspect of ecology, its place in environmental politics also demands the attention of historians of science. And this issue also provides an excellent opportunity to explore the social role of contemporary science at a time when the relation between knowledge and power is more contentious than ever. This book provides one perspective on this role, grounded in the history of ecology as it has unfolded in Great Britain, the United States, and Canada since the 1940s.

Several studies of the history of ecology have been written in recent years.

Reinforcing our understanding of science as an activity embedded in its social context, these studies have demonstrated how the ideas and values of ecology reflect society's inclinations to preserve, manage, or exploit nature. This book is distinctive, however, in both focus and method. As have other historians, I discuss how the concepts and methods of ecology (such as the ecosystem concept and the study of nutrient cycles) have changed over the past fifty years, and I note how these have been linked to social values, as expressed both through specific environmental concerns and through more general attitudes concerning the use and protection of nature. But I also interpret this history in light of the changing political role of scientific expertise. This role has been defined both by ecologists and by aspects of contemporary politics, including assumptions concerning the conduct of government and the organization and direction of science. And finally, I describe how these changes in science and politics, and their impact on ecology, can be understood in terms of the places in which ecologists work: their institutions, funding agencies, and managerial agencies to which they contribute their expertise. These institutional and political contexts have not often been explored in studies of the history of contemporary ecology.

As befits its subject matter, ecology is a complex discipline. Ecologists do not all share the same ambitions or respond to the same values and concerns. Once I had begun this project seven years ago, I realized that a realistic account had to respect this diversity, and the specificity of ecologists' experiences at different times and places. I have chosen to do this by telling not one story but four: one from Great Britain, two from the United States, and one from Canada. In recounting each episode I compare it with those discussed in earlier chapters. In chapter 8 I again compare these episodes to illustrate some more general aspects of the history of ecology and environmental politics.

The comparative method is not common in studies of the history of ecology (or of most other aspects of contemporary science, for that matter). This is understandable, given the difficulties inherent in comprehending even individual episodes, let alone several. However, I hope that my account, as have those of other historians, including Jonathan Harwood and Robert Kohler, will demonstrate the value of this method as a way of illustrating the contingency of scientific practice while building a better general understanding of science and its social role.

This book will be of most immediate interest to historians of biology or contemporary science. But I think it also will be useful to scholars and students who approach science with different concerns. Political scientists and historians have recognized science as important in the formation and implementation of environmental policy. In my own efforts to understand how science fits into the policy process, I have found their work helpful. But their studies, I believe, would also benefit from greater awareness of how scientific advice, far from being an independent input into policy, is itself shaped by its own history and by its institutional and political contexts. Finally, I have often noted the interest of many ecologists in the history of their discipline, and their role in society. I hope they will find this book helpful in their current efforts to navigate the rapidly evolving political context of ecology.

I've benefited from the advice and scholarship of numerous historians and scientists. I wish especially to thank Mary P. Winsor for her advice over the past decade. Samuel Hays, Trevor Levere, and an anonymous reviewer provided valuable comments on an early version of this book. Henry Regier has given much advice and encouragement.

Stanley Auerbach, F. Herbert Bormann, E. B. Worthington, Max Nicholson, Robert O'Neill, Gene Likens, Jack Christie, David Reichle, Richard Holmes, and Thomas Siccama generously took time to discuss their work and provide information. I thank them for helping to make my research a fascinating, often exhilarating experience. My thanks also to many historians who have provided me (through their published work) information, example, and inspiration, especially Charles Rosenberg, Sharon Kingsland, Arthur McEvoy, Donald Worster, Stephen Jay Gould, Ernst Mayr, and Daniel Kevles. Tangible support was provided by the Social Sciences and Humanities Research Council of Canada, in the form of doctoral and postdoctoral fellowships; by the Royal Society of Canada, for a year at the University of British Columbia; and by the Associates of the University of Toronto, for travel funding. Trent University provided congenial surroundings within which to complete this book.

I thank the Public Record Office, London, for permission to quote from several files, as listed in the references, the University of Toronto Archives for permission to quote from the Department of Zoology papers, and the Institute of Ecosystem Studies for permission to quote from the Hubbard Brook Ecosystem Study files. This book incorporates previously published material from *Science and Nature: Essays in the History of the Environmental Sciences* (Oxford: British Society for History of Science, 1993), 89–114; and the *Journal of the History of Biology* 28 (1995): 1–47. I thank the British Society for History of Science and Kluwer Academic Publishers for permission to use this material.

Above all, I wish to thank my family: my parents, for their encouragement; and Rita, Paul, and Emma, for their support, laughter, countryside rambles, and all other things.

Abbreviations

AEC   Atomic Energy Commission
ASLB   Atomic Safety and Licensing Board
BES   British Ecological Society
DOE   Department of Energy
EDFB   Eastern Deciduous Forest Biome
EPA   Environmental Protection Agency
ERDA   Energy Research and Development Administration
ESA   Ecological Society of America
ESD   Environmental Sciences Division
FRBC   Fisheries Research Board of Canada
HBES   Hubbard Brook Ecosystem Study
IBP   International Biological Program
NEPA   National Environmental Policy Act
NRIC   Nature Reserves Investigation Committee
NSF   National Science Foundation
ODLF   Ontario Department of Lands and Forests
OFRL   Ontario Fisheries Research Laboratory

ORNL   Oak Ridge National Laboratory
SCOL   Salmonid Communities in Oligotrophic Lakes Symposium
SCS   Soil Conservation Service
SFS   Society for Freedom in Science
SPNR   Society for the Promotion of Nature Reserves
SPOF   Strategic Plan for Ontario Fisheries
USFS   U.S. Forest Service
USGS   U.S. Geological Survey

1

Introduction: Ecologists, Institutions, and Society

Environmental issues have gained a permanent place on the political agenda. More than most political issues, they have also become associated with our scientific understanding of the world, especially that provided by ecologists. In North America and much of Europe, mass environmental concerns in the 1960s were said to indicate the arrival of an Age of Ecology, during which ecologists and writers inspired by ecology—Rachel Carson, Frank Fraser Darling, Barry Commoner, and others—would provide intellectual leadership for the environmental movement. Ecology has since been interpreted as a political movement, a source of moral values, or a guide for those wishing to live more gently on the earth.

But ecology is also a scientific discipline, with its own priorities and history. Ecological phenomena—the interactions of organisms with their environment—have been observed for centuries; as a distinct discipline ecology has existed for a little over a century. In the past fifty years it has expanded immensely. Within universities, governments, and other institutions, ecologists study the entire biosphere and its parts: a forested watershed or small lake, microorganisms within laboratory glassware, or populations existing only within a computer model. Ecologists have developed distinctive methods and theory while adapting techniques and ideas from other scientific fields. Professional associations, journals, and research programs testify to the growth of their institutions. Following a characteristic pattern of scientific disciplines, the maturation of ecology has strengthened ecologists' concern for the approval and esteem of their colleagues, displacing the more general standards of achievement in society. Thus, even as environmental politics has drawn on ecology, the discipline has developed an independent identity.

But ecologists have also had a recurring interest in environmental politics. Often they have stressed the technical utility of ecology in a world facing serious environmental problems.[1] Such ecologists as Paul Sears and Eugene Odum have also argued that ecology offers not only tools with which to address specific problems but also a novel, integrative perspective on the natural world, even a critique of human conduct. But some ecologists, noting that much funding for research on environmental problems has gone not to ecology but to other disciplines, have wondered whether ecology really is relevant to society's concerns. Others have resisted a prominent social role; one has even complained of "the word 'ecology' . . . being dragged back and forth . . . like a smelly red herring."[2] More substantively, evidence from ecology textbooks and elsewhere testifies to a distancing of the discipline from environmental concerns over the past decade. Among ecologists there evidently is no consensus concerning their place in environmental politics. Indeed, the debate is nearly as old as the discipline, and it has helped shape the history and structure of the discipline. The relation of ecology to environmental politics is also the focus of this book. I ask a specific question: What is the impact on a scientific discipline when it becomes a focus of society's concerns and values?

In recent years there has been much interest in the history of ecology and its role in environmental politics. In considering this role historians have often focused on certain underlying themes of ecological research: the significance of stability, or chaos, in nature; cooperation or competition within species; the relative importance of individuals, communities, and systems in the regulation of ecological phenomena; and the resilience and vulnerability of ecosystems to human impacts. The special task of ecologists, it has been suggested, has been to draw on these themes, to derive lessons concerning human conduct toward nature and each other; providing, in effect, intellectual leadership for the environmental movement, and sometimes for its opposition. It is these themes, therefore, and ecologists' willingness to act on them, that have shaped the role of ecologists in environmental politics. Historians have also emphasized the reciprocal influence of society, concluding that ecologists' theories, methods, and practical advice often mirror social values. Ecology, through its authority as scientific knowledge, has often legitimized society's attitudes, including predispositions to exploit, control, or preserve nature.

ECOLOGISTS AND THEIR INSTITUTIONS

This focus on underlying themes, and the values they imply, provides a valuable perspective on ecology and environmentalism. However, it also underrates the importance of where ecologists work and for what purposes, as determined by their institutions, funding agencies, and links with managerial or regulatory agencies. These institutional contexts mediate the relationship of ecologists with their surrounding society and are of central importance to their role in environmental politics. Accordingly, in this history I focus on ecologists' institutions and how they shape the choices ecologists make in their research. To demonstrate the significance of institutions, and their political contexts, I compare the development of ecology at four institutions in Great Britain, the United States, and Canada.

I begin with the Nature Conservancy of Great Britain. Its origins can be traced to the early years of World War II. With wartime activities accelerating the destruction of Britain's flora and fauna, and the government assuming a more active role in land-use planning, naturalists and ecologists mobilized to ensure protection of natural communities. Ecologists had other concerns as well. Led by Arthur Tansley, they recognized and skillfully exploited this opportunity to realize an elusive goal: establishing an institutional home for their discipline that was controlled by ecologists and accommodated the unique requirements of ecological research. In 1949 the conservancy became an autonomous research council.

During the 1950s the Nature Conservancy expanded rapidly, opening research stations throughout Britain. The conservancy also built and maintained a network of nature reserves and advised the government on conservation and land use. Ecologists' advice accommodated itself well to the preference for consultation and cooperation characteristic of British politics. Its management and advisory functions, and its independence from the immediate demands of government, also made the conservancy almost unique among ecological research institutions. I focus on the conservancy's Merlewood research station in the Lake District, where ecologists, led by J. Derek Ovington, studied woodlands ecology. They sought to understand the relation between soil conditions and trees, and the factors affecting the productivity of these ecosystems. They also experimented with techniques for the conservation of the English Highlands and nearby nature reserves.

I then turn to ecology in the United States, specifically research at the Oak Ridge National Laboratory in Tennessee. Begun in 1954 by Stanley Auerbach, the ecology program there grew almost continuously: by 1970 more than eighty ecologists and support staff members were engaged in a range of projects. This growth exemplified the role of the Atomic Energy Commission as a major supporter of ecology, including research on the movement and effects of radioactive contaminants in the environment, and the commission's view of ecology as consistent with its technological objectives, and with its self-defined role in American society. Other circumstances, including the ecologists' marginal status within a laboratory dominated by physical scientists, also shaped ecological research at Oak Ridge.

During the 1970s ecology at Oak Ridge continued to grow. By the end of the decade the Environmental Sciences Division had almost two hundred scientists and other staff members. The International Biological Program stimulated ecosystem research and modeling. Demands from other federal agencies for technical expertise—to solve environmental problems and to provide the means for choosing between environmental and economic priorities—also generated research. Studies of the environmental impact of nuclear power were of particular importance. Ecologists' involvement in studies of power stations on the Hudson River in New York illustrate the significance of regulatory demands to their work.

My account of American ecology continues with the Hubbard Brook Ecosystem Study in New Hampshire. This study began in 1963 as a collaboration between a forest ecologist, F. Herbert Bormann, and a limnologist, Gene Likens, with substantial support from Robert F. Pierce of the U.S. Forest Service. It has since grown considerably through the work of associates, visiting scientists, and graduate students, and the support of the National Science Foundation. The Hubbard Brook study has been largely the work of university scientists, often in collaboration with scientists of the Forest Service, which provided the experimental forest. This collaboration reflected an effort to combine methods and concepts from both ecology and forestry to build a distinctive approach to ecosystem research, based on the manipulation of entire watersheds. While the Hubbard Brook study is best known for researching the effects of acid precipitation, during its first fifteen years it focused more on the environmental impact of forestry practices. While ecologists' contacts with forestry research provided substantial benefits, in the 1970s the changing political context of American resource management practices, including forestry, obliged them to consider carefully their role in environmental policies.

Finally, I consider ecological research in Canada, specifically studies of fish populations and their environment conducted by the University of Toronto and the Ontario government. Beginning in the late 1940s university scientists and the provincial government formed a partnership to pursue research in order to provide a scientific basis for fisheries management practices in the Great Lakes and smaller inland lakes. This research addressed specific fisheries management priorities; it also helped justify a prevalent attitude toward resource use based on maximum exploitation. At the same time, this partnership provided opportunities for University of Toronto scientists to pursue theoretical questions concerning the dynamics of fish populations, physical and chemical limnology, and the environmental physiology of fish. In the early 1970s, however, the partnership became much weaker. The provincial government had developed its own fisheries expertise and relied less on university scientists. New university scientists also developed research addressing theoretical and empirical objectives relevant to other environmental and resource issues besides fisheries management. These looser ties between government and university, however, also provided, within a changing political environment, an opportunity for academic scientists to contribute to a new approach to fisheries management.

I had several motivations in selecting these institutions for study. One was their scientific importance. Within each country these institutions have been central to the formation of contemporary ecology. The Nature Conservancy was Britain's most influential organization for ecological research, helping to shape environmental management practices. Ecologists at Oak Ridge contributed several advances in theory, modeling, and research organization during the formative period of ecosystem research in the United States. The Hubbard Brook study has also been highly influential through innovations in watershed experiments, ideas concerning forest dynamics, and the response of ecosystems to disturbance. For many years Toronto was the most active center for aquatic ecology in Canada, training scientists for other institutions across the country, and guiding fisheries management practices of both the Ontario and federal governments. The development of ecology at each institution, therefore, constitutes a significant part of the history of ecology in each country. The varied approaches adopted by ecologists at these institutions also remind us that ecology is a remarkably heterogeneous discipline, encompassing a wealth of methods and research subjects. The range of its potential links with social values is just as broad. The choices that ecologists make from among these alternatives is influenced by their institutional contexts and their scientific training and expertise. Ecologists' own research ambitions are also important, as are their values: ecologists have often been strong environmental advocates. In these accounts of ecological research, the significance of those convictions and ambitions will be apparent.

These institutions also provide examples of different ways that scientific activity and practical responsibilities have been linked. The Nature Conservancy and the Oak Ridge Laboratory combined research and environmental management responsibilities within the same institution. However, the terms of this combination differed, reflecting, in part, the different practical tasks of the conservancy and the Atomic Energy Commission. At the conservancy, ecologists had considerable latitude in defining how their research would relate to practical concerns. At Oak Ridge, in contrast, physical scientists had the dominant role in defining this relation. At Hubbard Brook and at the University of Toronto, research was conducted within academic institutions separate from and yet associated with institutions with management responsibilities. Again, however, the specific terms of this association differed between these two institutions.

THE POLITICAL CONTEXT OF THE ENVIRONMENT

The institutions of ecological research and their relation to the concerns of society are themselves influenced by the assumptions and priorities that define political discourse and action—that is, by the political culture prevalent within a country. Comparison between institutions in these countries provides an opportunity to consider the significance of different political cultures to the social role of ecology. Three aspects of the political cultures of Britain, Canada, and the United States are especially relevant: environmental concerns (specific issues that are the substance of environmental politics), assumptions concerning the appropriate role of government in society, and views concerning the organization and direction of scientific activity. These aspects of political cultures vary by country and over time, affecting the role of ecology in society.

Ecologists at these institutions have addressed a range of environmental concerns: management of natural areas, forest harvesting, nuclear waste disposal, and depletion of fish stocks. Specific features of these concerns have helped determine which research approaches would be considered most relevant. Natural areas management demands study of the interactions between fauna and their plant habitat; forest harvesting implies the need to study the impact of vegetation removal; concern about the movement of contaminants from nuclear waste sites has justified study of the movement of elements in nature; depletion of fish stocks has led to research on the impact of removal of individuals on the dynamics of populations. It is evident, therefore, that specific concerns belong in a study of the history of ecology and its role in environmental politics.

Historians, however, have most often viewed such concerns not as significant in themselves but as clues to more basic human values. Specific concerns are only reflections, distorted by local circumstance, of a deeper essence. As Timothy O'Riordan has suggested, "Environmentalism . . . can no longer be identified simply with the desire to protect ecosystems or conserve resources—these are merely superficial manifestations of much more deeply-rooted values."[3] These values, as expressed by Gilbert White, Henry David Thoreau, John Muir, and other writers, and more recently by ecologists, have been considered a more appropriate focus of historical study than specific concerns.

The assumption generally implicit in analyses of environmental concerns as indicators of underlying values is that these concerns became widespread because certain individuals, or their ideas, became more widely known. The knowledge of the few—often grounded in an understanding of ecology—became the values of the many. Samuel Hays and others have argued, however, that this transformation and broadening of environmentalism was more the product of a transformation of society since World War II. Rising affluence has led North Americans and Europeans to seek not just the necessities and conveniences of life but also such amenities as clean air and water, which can be provided only by a protected environment. Greater leisure and mobility have also played a role by enabling more people to experience natural areas outside the city, thereby giving them a stake in the preservation of these areas. Concern about environmental quality, Hays concludes, stems "from a desire to improve personal, family, and community life."[4] This concern is less the product of the dissemination of the values of an elite than the result of the changing values and experiences of the people themselves.

This transformation also means that environmental concerns most often address the environment—not as it is understood by ecologists or others with special expertise but as it is experienced by humans generally. This may involve issues about which ecologists have expertise. For example, concern about maintaining and restoring natural ecological communities often demands extensive study of these communities by ecologists. In contrast, the increasing concern in the 1970s and 1980s about the impact of environmental contaminants on human health (not the health of other species) has demanded the expertise of such experts as epidemiologists and toxicologists. As an American environmental activist once noted, "We are about protecting people, not birds and bees."[5] It is about birds and bees and other fauna and flora that ecologists have expertise. Specific environmental concerns, then, actually can affect the place of ecology in society, pushing it toward the center or the periphery of environmental politics.

The range of assumptions concerning the appropriate role of government in society encompasses divergent answers to such questions as: Should government use its expertise to identify and seek goals for society, or should it merely provide an arena within which private interests can pursue their own goals? Should broad participation in decisions be encouraged, or should participation be limited to agencies and interests with official standing? The prevailing responses to these questions and others have often varied among Britain, the United States, and Canada, reflecting their distinct political traditions. The responses also have evolved over the past few decades, and they remain a focus of contemporary political debate. They are especially relevant to environmental politics because of their implications for the political role of scientific expertise. In areas ranging from natural resources management to the regulation of private activity to the development of technology, assumptions concerning the role of government have influenced views on how science should be applied to the problems that face society, and how the contribution of science should be evaluated. As the experience of ecologists demonstrates, the ability of ecologists to address these views and to fulfill the role expected of scientific expertise has varied.

Government funding of science in each of these countries has increased dramatically during the past fifty years. Ecologists have shared in this increase (if not to the same extent as such fields as physics or medical research). As a result, the ecological research communities in each country are now many times larger than they were in the late 1940s. This support has had several implications for the role of ecology in environmental politics. Most important, because much of this funding has been provided by national agencies, it has tied ecologists more closely to views concerning the organization and direction of scientific activity in each country—that is, to national science policies. In Canada, the United States, and Britain, these policies have evolved over the past few decades. Shifting from the view, prevalent in the 1950s and early 1960s, that the scientific community should determine its own priorities (subject to certain national objectives relating to security or development of technology), it has become increasingly accepted that science can be directed toward specific social or economic objectives. This evolution has affected the role of ecology in environmental politics by helping to establish whether this role is determined by ecologists or by nonecologists. The structure of ecological research institutions, and the links between ecology and environmental priorities that they are intended to foster, will also often reflect science policy priorities current at the time of their formation.

In examining the political context of ecology, I intend to bring together two traditionally separate areas of scholarship: history of science, and political science. The territory between these disciplines is potentially fertile but almost entirely unoccupied; its colonization should benefit both areas of study.

By considering various aspects of the political context of ecology we can build a more realistic portrayal of its role in environmental politics. This role is generated by the interaction of political culture, disciplinary priorities, and personal research ambitions. The arenas in which the interaction takes place are the institutions of ecological research. Ecologists' perceptions of the demands of society and of their colleagues—of where their research should go and what intellectual and material resources it can draw on to get there—are shaped by the demands and opportunities provided by institutions. It makes sense, therefore, to focus on institutions in a comparative history of recent ecology.

This account does not of course exhaust the variety of institutions of ecological research suitable for comparative study. For example, the Nature Conservancy has devoted itself to terrestrial ecology; British traditions in aquatic ecology or marine ecology may have different lessons to teach about ecology and environmental politics. Analogous comments may be made about ecology in Canada and the United States or in countries excluded from this comparison. Other subdisciplines of ecology that have maintained strong ties with societal concerns, such as human ecology, would also reward comparative study. Historical studies, this book included, have only begun to build an understanding of the place of ecology in contemporary society.

One feature shared by these institutions is their origin during or immediately after World War II. None existed before the war; five years after it ended, all were in operation. Some, including the Oak Ridge National Laboratory, which began as a component of the Manhattan Project, owed their existence to wartime initiatives. Others, including the National Science Foundation (which would eventually provide funding for the Hubbard Brook Ecosystem Study) and the research program of the Ontario Department of Lands and Forests, were the product of changing views concerning the relation between science, government, and society, stimulated in part by the demands of the war. These developments exemplified the transformation of the political and institutional contexts of ecology, and of the scientific community as a whole, since the 1940s.

The significance of the war to ecology is especially well illustrated in the origins of the Nature Conservancy. As I describe in the next chapter, a nation preoccupied with survival also provided the conditions that enabled ecologists to make their case for redefining ecology and conservation as responsibilities of government.

PART ONE

Great Britain

2

Origins of the Nature Conservancy

From a hill in Sussex, in the south of England, a diverse landscape of woodland and farmland extends to Chichester Cathedral and beyond to the seashore. Immediately below is Kingley Vale, a pocket of natural vegetation: a yew wood of "slow powerful trees that make their own cathedral-cool climate"; scrub and heath, with their characteristic and colorful flora; and a grassland rich in herbs and flowers. A Sarsen Stone, placed there in 1957, overlooks this view. Its inscription reads: "In the midst of this Nature Reserve which he brought into being this stone calls to memory Sir Arthur George Tansley, F. R. S., who during a long lifetime strove with success to widen the knowledge, to deepen the love and to safeguard the heritage of nature in the British Islands." Tansley was said to have considered this view the finest in Britain.[1] Like many British ecologists, he combined research with a deep affection for the countryside.

During World War II, naturalists, hikers, and others concerned about the British countryside debated the future of Kingley Vale and other natural areas. Tansley and other ecologists participated, insisting that ecological research was essential to maintaining these fragments of Britain's natural heritage. In 1949 a sympathetic government responded by creating the Nature Conservancy, an independent institution to support ecological research, provide advice concerning natural flora and fauna, and maintain nature reserves. For ecologists the conservancy was a milestone. While assuring the protection of many distinctive plant and animal communities, it also quickly became the leading ecological research organization in Britain, with several research centers and a program of grants to academic ecologists.

The Nature Conservancy, however, implied more than ecologists' success in protecting the countryside while gaining support for their research. The conservancy was established as British ecologists were attempting to redefine their discipline and to demonstrate that they could provide practical, authoritative advice on the management of the natural landscape. These objectives were reflected in new techniques of experimentation and survey, new theoretical perspectives, such as the ecosystem concept, and firm ideas on the relation between government and science. Convinced that their science could not flourish within existing academic or government institutions, ecologists decided to seek an institution dedicated to ecology. In the 1940s they assumed leadership of the nature reserves movement, making its objectives consistent with the theory and practice of ecological research and presenting reserves and nature conservation as an appropriate responsibility of government. Their success was apparent in the creation of the conservancy, which was designed according to the proposals of leading ecologists to fulfill the unique requirements of their research.

There were several distinctive features of the Nature Conservancy. Ecological research and opportunities for its application to environmental problems coexisted within the same institution. During the conservancy's first two decades, nature reserves, on which the conservancy focused much of its attention, attracted little public controversy. The conservancy also epitomizes a particular perspective on the natural world. Its leaders believed that the British countryside was the product of both nature and humans, and that its survival depended in many instances on humans' continuing to shape it. Perhaps most important, the conservancy was an institution largely created and led by ecologists. During the 1950s and 1960s ecologists themselves determined the objectives of conservancy research and how its results would be applied to practical concerns. All these circumstances influenced the development of ecological research at these institutions and its place within environmental politics.

In some respects the Nature Conservancy resembled the institutions in the United States and Canada that I discuss later. There were also major differences. In the following chapters I explore these similarities and differences, demonstrating the significance of the differing contexts of ecological research in Britain, the United States, and Canada.

APPLIED ECOLOGY

By the late 1930s, British ecologists could look back on several decades of efforts to obtain support for their discipline. Even in transforming the British Vegetation Committee into the British Ecological Society in 1912 they had been motivated, in part, by a belief that as a program with a wider scope, it would have less difficulty obtaining government funds.[2] This concern for funding was most apparent in discussions of the practical applications of ecological research. Until his death in 1955, Tansley was the most eminent British ecologist (fig. 1); he was also among the most persistent in demonstrating the utility of his discipline. Between 1924 and 1926 he coedited *Aims and Methods in the Study of Vegetation*, a book that explained how ecology could be ap-

Fig. 1. Arthur Tansley (courtesy of Hunt Institute for Botanical Documentation, Carnegie-Mellon University, Pittsburgh)

plied by foresters and agriculturists throughout the British Empire.[3] In 1927, in his inaugural lecture as professor of botany at Oxford, Tansley presented his vision of the future of botany, discussing at length its relevance to agriculture, forestry, and pasture management. Each area, he explained, depended on extensive understanding of plants and their environment. He also stressed that science itself would benefit from links to "practical human needs." He said, "It is perfectly hopeless to expect to develop or to maintain an efficient scientific department without the substantial monetary support that is rarely forthcoming unless a clear and unequivocal public utility other than a purely educational or academic utility can be demonstrated."[4] Practical relevance would also ensure support for graduates. As he noted several years later, "As soon as the various 'powers that be' have realized more fully that field ecology is the best of all training for forestry, for pastoral science, and for all careers in which a knowledge of the relation of vegetation to the land is of prime importance, the opportunities and facilities for such work will increase and multiply."[5]

But Tansley also argued that botany itself benefited from insights provided by practical problems. These problems were an "important part of the driving forces which are necessary for the fertile development of science, and if we try to divorce pure from applied science we drift almost inevitably into sterile academicism."[6] Tansley knew well the benefits that ecologists had derived from agricultural research. The work at Rothamstead Experimental Station, for example, had encouraged ecologists to study grasslands and the relations between plants and soil.[7]

But while benefiting from such contacts, ecologists were nevertheless unable to obtain from the Ministry of Agriculture much support for studies of strictly ecological problems. As R. George Stapledon complained in 1928, study of grassland communities affected by grazing was supported only when it could be justified by benefits to agriculture; research on vegetation of interest to ecologists alone would be possible only when ecology had "sufficient financial backing to found for itself a station unhampered by agricultural or other economic calls upon its methods of enquiry."[8] By 1939 only Charles Elton at Oxford had obtained significant support from the ministry for ecological research, for studies by his Bureau of Animal Population on rodents that eat grain. And this only came after a decade of modest, fairly unreliable grants from several organizations.[9]

In the 1940s Tansley led ecologists in an effort to demonstrate the relevance of their work to forestry. This culminated in July 1943 in a meeting of the British Ecological Society and the four British forestry societies, at which Tansley and other ecologists explained the assistance they could provide to forestry. The meeting had been precipitated, Tansley noted, by an article by

Sir Roy Robinson, chairman of the Forestry Commission, a government agency responsible for establishing and managing forest reserves. Robinson had explained that new forests were most successful when their planting and management followed the natural succession of woodlands. It was in particular necessary to be aware of which species could colonize bare land and which could replace pioneer species. An understanding of woodland ecology was therefore, Robinson believed, necessary to forestry management.[10] Robinson's comments apparently convinced Tansley that the Forestry Commission would be willing to provide opportunities for ecological research. He suggested, accordingly, that the commission permit the use of plantations as large-scale experiments. Ecologists should even be able to propose variations in forest planting so that these experiments could be as useful as possible. Both ecologists and the Forestry Commission would benefit from cooperation. Practical problems would provide starting points for research aimed at a general understanding of forest ecology; this understanding would then provide the basis for solution of specific problems.[11]

This effort to link ecology and forestry was especially timely because an increasing fraction of the countryside was being devoted to new forests. The month before the joint meeting the Forestry Commission had released its plan for postwar forestry, in which it proposed to create 5 million acres of intensively managed forests within fifty years. This would require new forest plantations on 3 million acres of land. Tansley was concerned about the aesthetic impact of these plantations and the replacement of deciduous woodlands by conifers:

> Young coniferous plantations, straight-edged and neat in form with precisely aligned saplings, their floors bare of the familiar woodland plants, introduce into the landscape an alien feature which cannot but offend the lover of the luxuriant irregular beauty of the countryside to which he is accustomed. And it is not only his aesthetic sense which is outraged. Plantations of conifers profoundly alter the plant and animal life of a region. Not only do they destroy the vegetation of open ground, but they change altogether and greatly impoverish the character of the woodland flora and fauna when the plantation is made on a woodland site.[12]

These plantations, Tansley argued, could cause soil deterioration and declining productivity. Such problems, he implied, demanded the attention of ecologists, who could also advise the Forestry Commission of areas suitable for reforestation, where their impact would be lessened. The highlands of north-west Britain, for example, while unsuitable for agriculture, could be used for "great forests of climax conifers developed and allowed to grow to maturity."[13]

Foresters were not unswayed by these arguments. The commission set up a few forest reserves and permitted ecologists to study its forests, even supporting research by a few of them. This support, however, had strict limits. This may have been because even though foresters did appreciate ecologists' insights into forest succession and soil deterioration they also had many concerns for which the expertise of ecologists was not immediately relevant, such as the effects of fire, wind, and disease, and methods of ensuring soundness of timber.[14] In general, the commission favored short-term research on problems encountered in planting new forests, not the long-term studies of established plantations that ecologists were most prepared to participate in. Another probable factor was doubt that foresters had much to learn from ecologists. Alex Watt, for example, argued that foresters were far ahead of ecologists in comprehending species patterns and interactions within forest communities. "Ecologists," Watt noted, "have a long way to go before they know enough about the relations between species in the plant community to make generalizations which will be useful in the practice of forestry."[15] Thus, in the 1940s support for ecological research from the Forestry Commission and from agricultural bodies remained a largely unfulfilled hope.

A paucity of support did not, however, prevent ecologists from discussing ambitious plans for research that required more support and facilities than they had ever had access to. At Oxford, for example, Elton was developing plans for his ecological survey of Wytham Estate, with the long-term aim of understanding the dynamic interactions between all the species of organisms living within it. Such a survey, Elton believed, was essential to bridging the gap between plant and animal ecology, permitting synthesis of "at least a working model of how whole ecosystems are arranged in nature, and how they work and interact."[16]

Tansley had coined the term *ecosystem* in 1935. The plants and fauna at any location, he had suggested, together with soil and climate, form an interacting ecosystem tending to equilibrium and able to resist, to some extent, disintegrative forces. Analyses of his ecosystem concept have tended to emphasize the context in which Tansley first presented the term: a critique of Clementsian ecology. Joel Hagen, for example, has described the concept as the ultimate result of a reworking of Frederic Clements's ideas into a philosophically more acceptable form: avoiding their excesses, particularly the view of plant communities as organisms (and the accompanying social implications of holistic organicism) while retaining their more useful aspects. Tansley's perspective on ecological succession, for example, drew on Clements's emphasis on dynamic processes. In revising Clements's ideas, Tansley, according to Hagen's valuable account, also applied the ideas of Hyman Levy, a

British mathematical engineer who viewed the universe in terms of partially overlapping systems.[17]

It is important to consider Tansley's ecosystem concept not only in the context of Clements's ideas but also in the context of the history of ecology in Britain. The ecosystem concept integrated the understanding that ecologists had gained of succession, the interaction between animal and plant communities, and the development of an equilibrium between the biota and its physical environment.[18] Ecologists in Britain, more so than elsewhere, had considered the interaction between plants and soil and the influence of grazing and other animal activities on plants. Tansley intended for the ecosystem concept to integrate this work into a unified perspective.

But the ecosystem concept could not only provide a perspective on previous ecology study, it also could suggest future research directions. It could help integrate separate studies of plant and animal ecology. It could provide the basis for new techniques, such as long-term studies of succession and field experiments, and for the adaptation of ideas from physics and chemistry, climatology and pedology, through which Tansley believed ecologists could develop a fully scientific perspective on nature. The ecosystem concept implied that study of an area must include all its aspects: plants and animals, because they affect each other; soil and climate, which affect the nature of the plant community; and the plant community itself, defined as a collection of species integrated by their interdependence.[19]

And as ecologists' "basic unit of nature," the ecosystem concept distinguished ecology from the study of both individual organisms and inorganic systems.[20] The universe, in Tansley's view, was made up of a vast number of overlapping physical systems, each tending toward a state of maturity characterized by equilibrium. The ecosystem was one of many such systems, but the one of special interest to ecologists. It provided a unified point of view, comprehending the "diverse features of the landscape with their progressive change or relative stability, the factors of climate and soil, whether these are wholly natural, as in virgin vegetation, or more or less profoundly modified by past or present human activity—and the living organisms themselves, the centre of the picture, as they are affected by the factors of their environment and by their interactions with one another. . . . It is the unifying point of view that makes ecology what it is."[21]

With these ideas and techniques emerging, some ecologists felt that their field was entering a phase in which new methods, concepts, and problems would reshape the discipline, shifting its orientation from describing the patterns of vegetation and animal populations to understanding the causes behind these patterns. Elton, for example, told ecologists that he was "absolutely convinced that we are on the threshold of a period when synthesis of animal and plant community facts and concepts will not only be possible but necessary, if we are to interpret ecological phenomena fully, and see them in the proper contexts."[22] Tansley agreed: "Ecologists have been largely occupied hitherto in describing the facts of natural succession, but we can have no thorough knowledge of a succession until we have discovered the actual effective factors which enable one kind of a tree to succeed in the presence of another but not alone. Only carefully designed experiments will discover the actual factors at work, and this is essentially the business of the ecologist."[23]

This attitude was also demonstrated in the reaction of leading ecologists to Tansley's *The British Islands and Their Vegetation* (1939), which summarized the work of British plant ecologists over the preceding four decades. William Pearsall found special significance in how this book "marks an epoch, the culminating point of a period in which the descriptive study of existing vegetation has been the leading interest of British ecologists."[24] According to Edward Salisbury, the book "may be said to mark the closing of the purely descriptive phase of the study of British vegetation. Future progress lies rather in the direction of causal ecology, of the study of the biological relations of species, and in the application of the principles of physiology and the factors of competition to the elucidation of the problems of plant and animal life under natural and semi-natural conditions."[25] Watt also saw Tansley's survey as a "landmark" indicating the end of one period of British ecology and the beginning of another: "It is the climax of a phase in which the main theme has been the relation of one plant community to another. The next stage, in my view, will be the intensive study and elucidation of the dynamic relations which hold between one species and another in the plant community itself."[26]

But this new stage depended on facilities that were not readily available. Flora and fauna of interest to researchers needed long-term protection. Tansley's research areas had twice been destroyed; he noted that "one of the most troublesome and irritating hindrances to ecological observations intended to serve as a basis for the study of successional change . . . is the liability to interference with or destruction of the vegetation of the area under observation by such events as clearing, felling, draining, gravel digging, change of ownership, or 'development.'"[27] Suitable areas were also needed for experimentation. Tansley noted in 1943 that for ecology a serious difficulty had been the "lack of sufficiently extensive opportunities for adequate ecological experiments, without which the science cannot progress."[28] Ecologists themselves also had to be assured of long-term support in order to complete intensive surveys or experiments. Secure employment and research funding, however, was not readily available. Tansley complained of the "difficulty of placing the young ecologist in a bread-winning job. Compared with the mycologist and even with the plant physiologist or geneticist he is still at a grave disadvantage."[29] He later recalled that he had even discouraged students from studying ecology because of the lack of opportunities.[30]

It was under these circumstances that a new occasion arose for demonstration of the practical benefits of ecology. This was generated by concern over the destruction of Britain's flora and fauna. To understand the approach that ecologists took to this issue we must turn to the long-standing interest in nature reserves.

PRESERVATION OR CONSERVATION

As early as 1873 it had been suggested that the government "preserve spots of primitive land-surface of which the vegetation was especially interesting."[31] In 1912, when Charles Rothschild founded the Society for the Promotion of Nature Reserves (SPNR), naturalists began a formal attempt to win support for the preservation of rare species and habitats through nature reserves. Rothschild and his colleagues at the National Trust and other preservationist societies were painfully aware of the gap between the few reserves they had acquired and the 273 areas "worthy of protection" that they had identified. They did not propose that the government take responsibility for reserves, only that it provide funds to voluntary societies that would select and protect reserves.[32] The concept of reserves did receive some official support. In 1931, for example, the Addison Committee, investigating the potential for national parks in Britain, recommended a system of national reserves and nature sanctuaries. The government, however, did not act, and as John Sheail has observed, nature preservation made little progress before World War II.[33]

Nature preservation assumed special urgency as the war deepened. Military ranges, camps, and airfields consumed land. Farmers, urged to produce as much food as possible, plowed under grasslands and heaths. Forests were cut down for timber. Taxes continued to force the breakup of large estates that had protected valuable areas. In addition, the government, preparing for postwar reconstruction, had begun developing mechanisms for planning and regulating development and land use. Lord Reith, the minister of Works and Buildings, initiated studies of industry, agriculture, and other factors affecting economic activities and land use.[34] One such study was by the Committee on Land Use in Rural Areas (the Scott Committee). The Forestry Commission and other government bodies also developed plans for the postwar development of the countryside.

The SPNR looked on these plans with some anxiety: "Many naturalists were concerned lest efforts to preserve the native flora and fauna for the benefit of posterity should be neglected."[35] In response, it campaigned to ensure a role for naturalists in the future of the countryside by arranging in 1941 the Conference on Nature Preservation in Post-War Reconstruction. One conclusion of the conference was that planning must protect flora and fauna useful to scientists through "strict" nature reserves in which "any work involving the alteration of the configuration of the soil or the character of the vegetation, any act likely to harm or disturb the fauna or flora, and the introduction of any species of fauna or flora . . . shall be strictly forbidden."[36]

Ecologists did not at this point play a major role in the nature reserves movement, possibly preferring to leave the initiative to amateur naturalists and recreational interests. In 1939, Tansley noted the efforts of the Commons and Footpaths Preservation Society and the National Trust to establish reserves. But he did not propose that ecologists become involved beyond perhaps surveying the plants and animals of trust properties. He apparently expected greater success to be achieved not through nature reserves but through the national parks movement, which "in some form seems likely to achieve a positive result in the immediate future."[37] Thus, although the British Ecological Society was one of fifteen organizations taking part in the 1941 Conference on Nature Preservation, the meeting was led by Herbert Smith, an expert on gems and the honorary secretary of the SPNR.

By 1942, however, ecologists had decided to play a more active role in the nature reserves movement. Their decision may have been motivated by the interest the government was demonstrating in the issue. That year representatives of the Conference on Nature Preservation met with William Jowitt, chairman of the ministerial committee concerned with reconstruction.[38] Jowitt invited them to form a committee to advise the government on reserves. This group, the Nature Reserves Investigation Committee (NRIC), was also led by Smith, but it included several ecologists, including Cyril Diver, Pearsall, and Salisbury. In its 1943 report the committee presented a list of proposed nature reserves and advised that the sites be subject to a "rigid and continuing control under scientific direction." This control would be the responsibility of experts on plant and animal communities employed by a "National Reserves Authority" that would be a department of the Ministry of Town and Country Planning.[39]

The ecologists had selected many of the reserves proposed by the NRIC. But their influence was especially evident in the committee's emphasis on active management of reserves. By this time ecologists had concluded that most British plant and animal communities had been affected by human activities. British woodlands, for example, were intermediate between virgin forest and plantations, exhibiting aspects of both the primeval forest and a long history of care and exploitation.[40] Grasslands were also, as Harry Godwin noted in 1929, shaped by human uses. When they were subjected to continued pasturage or cutting their succession could be deflected into a different but nonetheless stable climax community. If management ceased, succession might take a different course, eliminating the existing community. Through such insights and observations the effects of human activities had become an integral part of ecologists' theoretical understanding of plant communities. In defining his conception of ecology, Tansley stressed in 1935 that however overwhelming is the human impact, it remains a factor that ecologists must consider: "Regarded as an exceptionally powerful biotic factor which increasingly upsets the equilibrium of preexisting ecosystems and eventually destroys them, at the same time forming new ones of very different nature, human activity finds its proper place in ecology." He reminded his colleagues, "We cannot confine ourselves to the so-called 'natural' entities and ignore the processes and expressions of vegetation now so abundantly provided us by the activities of man."[41] Practical experience with the sequence of events following designation of a nature reserve had also illustrated the need for human intervention to maintain preferred communities. At Woodwalton Fen the cessation of grazing, haymaking, and peat-cutting had led to the invasion of bushes to such an extent that the reserve quickly became an impenetrable thicket. Wicken Fen nature reserve had to be periodically cleared of bushes to maintain its characteristic vegetation.[42]

In the NRIC report, as well as in a 1943 report of the British Ecological Society, ecologists insisted that nature reserves be managed so that their distinctive plant and animal communities are maintained. They called the management of natural areas *conservation,* a word with a long history in Britain.[43] Ecologists' use of the term, however, was likely intended not so much to evoke this history as its usage in the United States and Canada, where it had signified the scientific management of natural resources since the turn of the century. By the 1940s, British ecologists held North American conservation in high regard. Elton praised the "fine tradition of the United States Federal wild life services" in his review of *Natural Principles of Land Use,* by Edward Graham, the chief of the biology division of the U.S. Soil Conservation Service.[44] In 1941–42, Julian Huxley visited the Tennessee Valley Authority, which he saw as the application of conservation and planning principles to an entire region. In adopting the Americans' terminology British ecologists evidently hoped to emulate the success of North American conservationists in achieving recognition and support for their work. Their willingness to learn from the North American example was also suggested by the work of the Wild Life

Conservation Special Committee. I discuss this committee below; it is appropriate to note here, however, that while developing a proposal for a nature conservation institution the committee studied the work and organization of the U.S. National Parks Service and the Fish and Wildlife Service.[45]

The insistence of ecologists on the necessity of scientific management of reserves had several implications. Most important, it established the ecologists as the authorities best able to manage these reserves. Management required an understanding of the dynamic processes of natural communities, especially their successional development. Such an understanding had been accumulated by ecologists over the preceding four decades. Thus, as nature reserves moved closer to becoming a government responsibility, this authority brought ecologists closer to official recognition of the practical utility of their expertise. Effectively, ecologists sought to define nature reserve management as a professional activity. In doing so they excluded others from this activity, particularly the amateur naturalists who had been a driving force in the nature reserves movement. By designating their research and its application in management as conservation, ecologists distinguished themselves from amateur naturalists and their demands for the "preservation" of nature. Reliance on preservation, Tansley insisted, demonstrated an ignorance of ecology: "Some nature lovers who have little actual acquaintance with the ways of nature and her reactions to human activities constantly advocate a policy of 'letting nature alone' in nature reserves. This is only justified in the few places where there has been little or no human interference. Over most of the country the natural vegetation . . . has been decisively altered and more or less stabilised by man, has become in fact what we call 'semi-natural,' and if the human regime is discontinued it will change into something else."[46]

The notion that intervention was necessary to maintain plant and animal communities was useful for ecologists in another way: it helped dissolve the distinction between natural and artificial communities. There was very little, if any, pristine nature left—only plant and animal communities affected to varying degrees by human activities. Most vegetation in fact was neither natural nor artificial but "semi-natural." Knowledge gained through study of manipulated communities could therefore provide insights into communities once considered "natural" but that in fact differed only in degree of human influence. This conclusion gave more credibility and authority to ecologists developing experiments, studying forest plantations, or arguing for the relevance of ecology to agriculture and forestry management.

Another advantage to ecologists of the notion of conservation was its compatibility with the government's intentions for postwar planning. Just as ecologists were arguing that management was necessary to maintaining valued features of Britain's flora and fauna, government planners and other experts were arriving at a similar conclusion for the countryside as a whole. Patrick Abercrombie, Thomas Sharp, and others had often argued that the British countryside always had been shaped by a dynamic interaction between human activities and natural features. This perspective achieved some official status in 1942, in the influential report of the Scott Committee:

> The landscape of England and Wales is a striking example of the interdependence between the satisfaction of man's material wants and the creation of beauty. The pattern and the beauty of the countryside as we know them to-day are largely the work of man during the past few centuries. Its present appearance is not by any means entirely the work of nature and it is not enduring, for nature is dynamic, not static. Experience has shown how quickly land can revert to an unkempt, wild and ragged condition, even when it is only neglected and not wholly abandoned. The beauty and pattern of the countryside are the direct result of the cultivation of the soil and there is no antagonism between use and beauty.[47]

The countryside, the committee concluded, "cannot be 'preserved.' . . . It must be farmed if it is to retain those features which give it distinctive charm and character."[48] This view, it is evident, was consistent with the case made by Tansley and other ecologists that nature reserves should be not merely preserved but also managed. By arguing that they could guide reserve management and conservation they effectively presented their expertise as essential to and consistent with government plans for the countryside as a whole.[49]

For ecologists, nature reserves were only one element in a broader program of conservation research and practice. Although several ecologists were members of the NRIC, they decided in 1942 to prepare their own report on nature conservation. They did so because they felt that the "interests of ecology as such and the views of ecologists ought to be independently formulated."[50] The following year the report of the British Ecological Society (drafted by Tansley) endorsed the NRIC's proposal of a system of reserves that would be managed by experts and used for ecological research and other purposes. But the society also recommended a "National Wildlife Service," to conduct ecological research. This service would not be within the Ministry of Town and Country Planning but would be an independent body under the Privy Council, comparable to the Medical Research Council. Ecologists feared that if the Ministry of Town and Country Planning became responsible for nature reserves, as the NRIC had proposed, then independent research would be impossible. Such an arrangement would also hinder the application of research results to other responsibilities of government, including agriculture and forestry. Only an independent entity would be able to "create its own policy and traditions," they believed.[51] These ideas were developed further in a second publication of the British Ecological Society, released in 1945 as a "Memorandum on Wild Life Conservation and Ecological Research from the National Standpoint." It argued for an ecological research council with independence and powers similar to those of the other research councils; it would be composed of at least one institute of terrestrial ecology and a biological service responsible for managing nature reserves and advising on the conservation of plants and animals.[52]

Ecologists expended considerable energy preparing their proposal for an ecological research council. Tansley even wrote a small book explaining to a wider audience the merits of nature conservation.[53] They expected that such a council would fulfill their long-standing aim of a secure institution for ecology while demonstrating its relevance to practical problems. But during the 1940s they also devoted much attention to the apparently unrelated issue of the autonomy of science. Many ecologists, led by Tansley, believed strongly that scientists had to be free to set their own research priorities. To understand the role of this belief in the events leading up to the establishment of the Nature Conservancy it is necessary to examine the contemporaneous debate over relations between science, government, and society.

FREEDOM IN SCIENCE

Economic depression, World War II, and the status of science under fascist and Soviet governments provoked debate in Britain over the place of science in society. During the 1930s several scientists and historians argued that social and economic conditions, not curiosity, drive science, and that the power of science to transform society must be recognized and controlled. In 1931 Soviet historians visiting London insisted that the division between pure science and society was an illusion fostered by class structures and that science should be planned in order to be of maximum benefit to society. In 1939 John Desmond Bernal argued, in *The Social Function of Science*, that only through rational organization and planning, as in the Soviet Union, could science avoid waste and inefficiency and provide genuine social and economic benefits.[54] Such views had by then become widely accepted, even by such leading scientific organizations as the Association of Scientific Workers and the British Association for the Advancement of Science. Between 1932 and 1945 this Social Relations of Science Movement exerted great influence within the British scientific community.[55]

But for one Oxford zoologist Bernal's book provided both a stimulus and a target for counterattack. In November 1940 John Baker sent to forty-nine

British scientists a statement expressing concern that an "influential group of scientists" believed that researchers should be directed to study utilitarian problems and to work within organized groups. According to Baker these views endangered both pure science as an activity pursued as an end in itself, and individual research.[56] Many scientists answered his call, and together they formed the Society for Freedom in Science (SFS) to unite opposition to the direction of science toward social priorities. They argued that even in the absence of immediate practical benefits, science, like literature and art, was worthwhile in itself; that discovery depends on unrestricted curiosity; that no planner could predict which areas of science would produce new discoveries; and that, therefore, efforts to plan or control science would stifle creativity.

Tansley was active in the SFS, sharing its leadership with Baker and Michael Polanyi. Tansley helped draft the society's statement of beliefs, was on its executive committee, served as its first acting chairman and later was vice-president. His 1942 Herbert Spencer Lecture at Oxford was an eloquent statement of the SFS's ideals.[57] He remained active in the SFS until shortly before his death in 1955. Baker wrote then of Tansley, "He was always a good counsellor, urging restraint when wild schemes would have damaged our cause, energetically supporting strong action whenever it could be successful. He was an eminently wise man, thoughtful, full of experience, moderate, optimistic. . . . Whenever it was suggested that opinion was swinging in our favour and that we might relax our efforts, or even wind up the Society, he objected vigorously."[58] While the social relations of science movement provided the immediate impetus to the SFS, Tansley had long considered freedom essential to science. In his 1927 lecture at Oxford he reminded his audience that an "essential condition of successful development is freedom from utilitarian *compulsion*." Further, "good effective research cannot be done to order, nor can the spirit of investigation be manufactured artificially. That spirit depends absolutely on the driving force of natural curiosity."[59]

Tansley, however, was not the only ecologist in the SFS. More than a quarter of the society's members—seventeen out of sixty-one—were ecologists.[60] This must reflect, in part, Tansley's selection of members from among his own colleagues. Nevertheless, that the number of ecologists in the SFS was disproportionate to their significance within British science requires a stronger explanation—particularly because one might expect ecologists to have had little cause to be concerned about utilitarian compulsion. Because their research was considered to have limited practical significance and because it received little funding, government would be less likely to direct it. In a sense, indifference could ensure ecology's independence.

But it could also ensure ecologists' poverty. In the freedom in science movement Tansley was as concerned about funding as about freedom; indeed, the two were inseparable: "Pure science is necessarily the basis of the whole scientific fabric, because the original discoveries on which it is founded are unforeseeable, and cannot be provided to order or planned in advance; and it is for this reason that pure science, under the conditions of modern life, must . . . have money at its disposal which can be used by the people who spend their lives in making investigations in accordance with their own interests. They must have their freedom and they must have money; and that money must not be controlled by any outside authority which is activated by motives of practical value."[61] Tansley and his colleagues were, as they had been for many years, eager to arrange secure support for ecology. Bernal and the social relations of science movement, however, threatened progress toward this goal. While ecologists had argued that their research had practical benefits, their lack of success in obtaining support indicated that few within government had been convinced. If, then, science was to be directed toward immediate economic or social concerns, ecology would inevitably be neglected. This, of course, had been the case before the war, when government support for forestry and agriculture research had provided only meager opportunities for ecology.

The danger that the social relations of science movement posed to ecology heightened as ecologists tied their fortunes to nature conservation and its explicitly noneconomic objective of the protection of the British natural heritage. Tansley wrote in 1945: "If we keep and manage [natural areas] properly we can hand down to our descendants a Britain which is not only economically prosperous, but retains something of its distinctive natural historic character, continues to delight our sense of beauty, and affords rich material for the increase of knowledge."[62] The British landscape, Tansley insisted, had a value that transcended economic utility. And so did ecology, because it could provide the scientific basis for the care of this landscape. Baker made the same point in describing Tansley's massive *British Islands and Their Vegetation* as of no immediate practical application but of great cultural value.[63] If support for science was determined on the basis of immediate social and economic priorities, these more intangible benefits of nature conservation and related ecological research would perhaps be ignored.

Besides advocating autonomy and funding for science, Tansley also ensured that the principles of the Society for Freedom in Science would be compatible with ecology. Baker had originally distinguished between organized groups working on projects of economic benefit and individuals studying pure science. Pure science, he believed, was best performed by individuals. But when Tansley, Polanyi, and Baker established the principles of the society they concluded that researchers should be free to "work separately or in collaboration as they may prefer. Controlled team-work, essential for some problems, is out of place in others."[64] Tansley himself noted that some problems in pure science could be best approached through teamwork.[65] He was evidently responsible (with Polanyi) for the Society for Freedom in Science not being opposed to organized activity in science. Tansley may have taken this position because he believed that reliance on individual research could hinder ecosystem study. Because most ecologists were specialists, an individual could not do a complete study of an ecosystem. It required "collaboration between an animal and a plant ecologist, or by a team. . . . Such collaboration, to be really successful, must be genuinely joint work in which each partner fully follows and understands the contributions of the others. Only thus can a unified picture of the whole be built up."[66]

This distinction between individual and group research reminds us that scientific autonomy can mean different things: individual autonomy (each scientist determining his or her own research objectives) or disciplinary autonomy (objectives being established by scientists within a discipline, who may then elect to achieve them through forms of research organization in which some scientists provide direction to others).

These concerns also suggest a need to reinterpret the controversy over freedom and planning in science. William McGucken has described this controversy as the opposing views of the social relations of science movement and the Society for Freedom in Science, and he argued that the controversy ended when the government reaffirmed the autonomy of the scientific community.[67] But for ecologists autonomy was not the sole issue. They were also concerned about the demand of the social relations of science movement that science be directed toward social and economic priorities, and the threat that this demand posed to support for ecological research. This controversy, therefore, was in part the consequence of economic and disciplinary concerns, not simply the result of differences in ideological commitment.

Tansley led ecologists in efforts to defend scientific autonomy and to gain an institution for ecological research. As the most eminent British ecologist, a leader of the Society for Freedom in Science and the drafter of the British Ecological Society's report on nature conservation, he used the issue of nature reserves and their management to demonstrate the practical implications of scientific autonomy. Other ecologists, particularly Elton, shared Tansley's conviction that ecological research must be independent.[68] Led by Tansley, they offered a detailed argument for establishing an autonomous institution for ecological research.

Their insistence on this new institution, to be independent of the Ministry of Town and Country Planning, put them at odds with the Nature Reserves Investigation Committee and with John Dower, who was then shaping government policy on national parks. In 1945 Dower recommended formation of a Wild Life Conservation Council, which would provide scientific advice on the management of nature reserves and national parks. This council would be within a National Parks Commission and would report to the minister of Town and Country Planning.[69] Dower's strong opposition to the ecologists' proposal that nature reserves be independent of the National Parks Commission reflected a lack of consensus on the appropriate organization for reserves and ecological research. Besides Dower and other Ministry of Town and Country Planning officials, the Agricultural Research Council and the Treasury also opposed forming an independent institution for nature conservation.

That year the minister of town and country planning, after the cabinet refused him permission to establish a National Parks Commission, appointed a committee under Sir Arthur Hobhouse to review the issue. The Wild Life Conservation Special Committee was also established because the minister wanted "to have the proposals of the Dower Report on Nature Reserves and Wild Life Conservation examined by a small group of experts," which could then recommend a response.[70] The committee, chaired first by Julian Huxley and then by Tansley, was composed primarily of leading ecologists selected by Huxley. It was charged with determining which organization should be responsible for nature conservation. Among committee members Tansley was far from alone in his beliefs regarding ecological research and its organization and its relevance to conservation. Diver had prepared large parts of the memoranda of the Nature Reserves Investigation Committee, and Elton had served on the British Ecological Society's committee on nature conservation.[71] Elton's recent Bureau of Animal Population work on rodent control had also convinced him that ecological research would be best fostered by an independent research council. Another member was the ornithologist Max Nicholson. In his first book, *Birds in England* (1926), the twenty-two-year-old Nicholson had presented a highly original critique of both the depredations of bird and egg collectors and the existing policies of bird protection. Both, he reasoned, should be replaced by a focus on the "balance" of birds and their habitats—that is, their ecology. Since publishing his book he had been involved in the Oxford Bird Census, the Edward Grey Institute of Field Ornithology, and the British Trust for Ornithology. Each reflected his ambition to develop a more systematic organization of the efforts of amateur ornithologists.[72] Reflecting his status as a cofounder of modern British field ornithology, he had been chosen to represent the ornithological perspective on the Wild Life Conservation Special Committee. He would also make an essential contribution to the practical success of the committee. The committee itself, although officially reporting to the Hobhouse committee, acted independently.

The appointment of ecologists to this expert committee indicates that ecologists had taken the place formerly occupied by amateur naturalists as the government's advisers on nature reserves. This, and the bypassing of the NRIC, which had been established three years before to provide just such advice, was at first not well received. When J. B. Bowers, secretary of the Hobhouse committee, requested from Herbert Smith the NRIC's materials, Smith answered curtly, "I may remind you that the . . . [NRIC] were appointed . . . to advise government departments on matters relating to nature reserves and natural history generally. It would appear as if your ministry had ignored, or at least overlooked, the existence of the committee."[73]

Throughout the two years of its work the wild life committee sought to reconcile two objectives: the preference of Tansley and other ecologists for an independent ecological research council under the Lord President, and the committee's official duty of advising the Ministry of Town and Country Planning. In its first attempt at a compromise the committee proposed that a wild life conservation advisory body or research council, reporting to the Lord President, be responsible for research, with reserve management the task of a Wild Life Conservation Administrative Body within a National Parks and Wild Life Conservation Authority, reporting to the minister of town and country planning.[74] The committee presented this as an interim proposal in November 1945 with the proviso that it would require close ties between research and management: "If the primary purpose of [nature reserves] is to provide field laboratories for scientific research, and if Wild Life Conservation policy is to be framed on the best available scientific advice, it will be essential that the Wild Life Conservation Commission should be closely integrated with a National Research organization that is or may be set up in the Biological or Ecological field."[75]

The committee, however, quickly came to consider this compromise unacceptable. Placing reserve management within one government body and research within another would make their coordination much more difficult. And as the committee explored the implications of conservation, the necessity for such coordination became ever more evident.[76] Reserves could be fully useful for research only if scientists were involved in all aspects of their management. This was especially important because of the growing interest among ecologists in experimental research. In February 1946 committee member John S. L. Gilmour described to his colleagues the opportunities that reserves presented for such botanical experiments as studies on the growth of plants outside their normal range and on the effects of exclosure of rabbits and other herbivores from plant communities. Two months later Nicholson presented the case for experimental bird reserves, arguing that such reserves would enable studies of the effects of environmental changes on birds and help to train ecologists who would be needed by nature conservation and research bodies in the future. Reserves were, as Huxley expressed it, the "essential laboratories" of the nature conservation body.[77] It was also appropriate that they be administered separately from national parks because while parks were to be for public use, access to reserves would be restricted in order to protect the flora and fauna and prevent interference in research.

By February 1946 the committee had, accordingly, decided not to compromise on its preference for an independent body for reserves and ecological research. Just as the British Ecological Society had done two years before, it chose to recommend that both management and research within nature reserves be recognized as a matter of science, not planning, and that it be the task of a single body reporting to the Lord President, not the Ministry of Town and Country Planning. A biological service would manage the reserves, conduct related research, and respond to requests from other agencies for advice on conservation. In addition, one or more "terrestrial research institutes" would be dedicated to "fundamental and generalized" ecological research. Entirely separate from the biological service, these institutes would ensure the independence of the researchers.[78]

This decision to depart from the terms of its appointment by the minister of town and country planning and recommend a scientific body under the Lord President was a product of the committee's perception of what had become politically possible. Immediately after the war, science was enjoying unprecedented prestige. The scientific community and government then agreed that science, having helped assure victory, should be permitted to set its own priorities. The Lord President (Herbert Morrison) told scientists in 1946 that "science and scientists must have an independent life and an independent existence of their own," although government could guide the application of new discoveries.[79] The committee therefore had a greater likelihood of having its proposals accepted, if conservation and ecological research were defined as a strictly scientific matter.

The wild life committee had even been publicly advised to link conservation with science. An editorial in *Nature* had suggested that "the word 'nature' has come to be associated . . . with a somewhat childish or eccentric form of botanizing, bird-loving and butterfly-hunting. If the more fashionable word 'science' could replace 'nature' on the subcommittees' note-paper they would probably find the task of co-operating with local authorities less uphill work."[80] Ecologists already knew the value of such a strategy. Elton had told the committee that the political difficulties in acquiring nature reserves "might be mitigated by calling them 'scientific nature reserves,' thus using the present momentous prestige of 'the scientist,' rather than the slightly shabby reputation of 'the naturalist.'"[81] The committee evidently recognized that its recommendations would carry more weight if cast in terms of science. But if an independent nature conservation body was politically realistic, to argue for a full research council similar to the Agricultural Research Council and the Medical Research Council was not. Committee members felt it "inadvisable to ask for everything at once, and [thought] that they were more likely to succeed with a modest demand. They hoped, however, that both grant-giving powers and full status as a Research Council would in the long run accrue to the proposed advisory board."[82]

A further advantage of defining nature conservation as a scientific task and therefore necessarily independent of a National Parks Commission was that this proposed commission had begun to encounter political opposition. Even as the Hobhouse and Huxley committees conducted their inquiries a plan was being drafted to give county councils broad authority over land use and development. This resulted in the Town and Country Planning Bill, which received royal assent in August 1947. With their newly acquired authority over planning these councils now viewed the proposed National Parks Commission as a potential rival. Their opposition to the commission obliged Lewis Silkin, the minister of town and country planning, to limit its role within the land use planning process to providing advice to the councils.[83] By proposing that nature conservation be independent of the National Parks Commission, therefore, the ecologists could lessen the risk that opposition to national parks would extend to nature reserves.

While preparing its report, the wild life committee took steps to ensure its acceptance by the government by meeting with the Cabinet's Scientific Advisory Committee, in part to gain its support for the recommendations. They especially wanted the advisory committee to agree that conservation be the responsibility of the Lord President, because the government's acceptance of this recommendation was by no means inevitable. Proposing new responsibilities for the Lord President when the committee had been created by another minister was obviously a situation of some "delicacy."[84] In addition, Dower, still influential within the Ministry of Town and Country Planning, remained entirely opposed to the separation of national parks from nature reserves. The committee hoped for support from the advisory committee "in order that the National Parks Committee might be satisfied that the Wild Life Committee had a satisfactory alternative to the Dower scheme."[85]

The advisory committee had already begun considering the nature reserves question. Prompted by a report from the British Ecological Society and by one from the Royal Society, the committee invited John Fryer, secretary of the Agricultural Research Council and a member of the advisory committee, to summarize the issues involved. According to John Sheail, Fryer's note formed the basis for a report by the advisory committee to the Lord President. The report advocated a system of nature reserves, one or more institutes of terrestrial ecology, and a scientific service to manage the reserves and provide advice on conservation.[86] The wild life committee, by meeting with the advisory committee and sending it at least one memorandum, may have also helped persuade it to make such a proposal. This is suggested by the advisory committee's adoption of the wild life committee's advice to recommend that the conservation body be associated with the Agricultural Research Council. It is at least clear that the advisory committee's proposal, being similar to the one prepared by the wild life committee, gave it considerable credibility.

In July 1947 the wild life committee presented its report, which recommended establishing nature reserves and conservation areas; a biological service to manage these reserves, conduct research, and provide advice; and four terrestrial research institutes for fundamental studies. The report summarized the deliberations of the committee over the previous two years. Two general principles formed the basis of its recommendations. The first was that conservation was preeminently a matter of science. While the conservation of nature was, as the committee noted, a difficult problem in a crowded, highly industrialized nation, members confidently asserted that "the solution of the problem can be expressed . . . in three words: Research and Experiment."[87]

The second principle was that the proposed Nature Conservation Board must be independent. The committee explicitly recommended that it not be attached to a national parks authority, to the Ministry of Town and Country Planning, or to any other government department. The preferred option was for the board to be a research council under the Lord President. However, the committee took the more politically realistic route of advocating that the board be attached to the Agricultural Research Council (ARC), from which it did not anticipate interference in the board. Within the board, the terrestrial research institutes would also be independent.

A striking aspect of the wild life committee's recommendations was that they addressed difficulties that had hindered ecologists and their research. Most obviously, a biological service and research institute could employ many ecologists. In addition, an independent Nature Conservation Board would enable ecologists to avoid the problems encountered in obtaining support from bodies, such as the ARC or the Forestry Commission, whose concerns were not always compatible with ecological research. Research sites protected from disturbance or destruction were equally important. Nature reserves provided many such sites, as well as examples of every type of plant and animal community of interest to ecologists.

The wild life committee's recommendations also solved problems that the academic world posed for ecologists. Resistance from universities and other biologists was for many ecologists a familiar story, from Tansley's complaint in 1914 that ecology "occupies a very small, in many cases almost a negligible, place in our University curricula" to Cambridge ecologist Harry Godwin's recollection of being told by "more than one leading botanist" in the 1920s that ecology had no future to Elton's difficulties in obtaining support for his Bureau of Animal Population at Oxford in the 1930s.[88] Another problem was the difficulty in arranging cooperative studies by scientists from several disciplines and university departments. Tansley and other ecologists considered such cooperation essential to ecosystem studies. But the academic division between botany and zoology reinforced, as Elton complained, the "split between the plant and animal ecologists' outlook," hindering an integrated perspective on ecosystems.[89] Finally, academic duties presented a problem. Tansley noted that a "minor but quite real academic drawback is the frequent necessity of spending long working periods in the field, which does not fit in well with existing academic regulations."[90]

Each of these problems could be solved by the wild life committee's proposal for independent, multidisciplinary research institutions, working within protected and expertly managed nature reserves. If enacted, the proposal would make possible the innovations suggested by leading ecologists: in techniques, including a greater use of experimentation and intensive survey; in new concepts, especially the ecosystem concept; and in the replacement of descriptive studies by a focus on the causal relations behind observed patterns in nature.

The wild life committee's proposals were strikingly similar to Tansley's ideas on the organization of science. Both Tansley and the committee insisted that pure research and its application could profit from interaction, but only if the appointment of researchers, selection of research projects, and allocation of funds remained in scientific hands. Under Tansley's chairmanship, the report of the wild life committee became a practical expression of Tansley's views on the organization of science. The committee's members were certainly prepared to follow his views: most were ecologists, well aware of the needs of ecology and nature conservation, and respected Tansley as the most distinguished member of their discipline. They also shared his views on scientific autonomy; of the ten committee members, six were also members of the Society for Freedom in Science.

In January 1948, Morrison, as Lord President, accepted responsibility for acting on the wild life committee's recommendations. On April 29 he told Parliament that the government had accepted its proposal for a Nature Conservation Board and Biological Service, to be associated with the ARC. This acceptance had likely been facilitated by the committee's meetings with the Cabinet's Scientific Advisory Committee: Morrison told Parliament that the case for a conservancy had become very strong when several independent sets of experts agreed on its merit. After, as Nicholson described it, these plans were "steered gently through mechanisms which would normally have quietly buried them," in February 1949 Morrison announced that the Conservation Board and Biological Service would be united within a single body, to be called the Nature Conservancy. Accorded research, advisory, and conservation duties by Royal Charter in March 1949, and statutory power to designate nature reserves by the National Parks and Access to the Countryside Bill of 1949, the conservancy became an independent research council similar to the ARC and the Medical Research Council, but it also held powers of land tenure and control.[91]

Several factors abetted the government's rapid adoption of the wild life committee's recommendations. As I have noted, the recommendations were considered separately from the proposals for national parks and therefore were not subjected to opposition from local governments eager to protect their own planning authority. Nicholson also played a central role in their adoption. As a member of the wild life committee and as head of the office of the Lord President, he contributed to both the report and the formulation of the government's response to it. Perhaps most important, however, was the prestige of postwar science and the interest in applying science to government responsibilities. The postwar Labour government increased support for civil science from £6.5 million in 1945 to £30 million in 1950. It also arranged new mechanisms, such as the Advisory Council on Scientific Policy, to provide it with scientific advice.[92] The importance that the government attached to such advice was reflected in the Lord President's explanation to the Cabinet of the need for access to expertise on nature conservation: "The Government is constantly taking action liable permanently to affect the fauna, flora and even the geography of the country without having at its disposal any channel of authoritative scientific advice about the probable results of its action, such as is available in all other fields of natural science."[93] Because they had stressed that conservation was a scientific matter to be approached through "research and experiment," the new authority of science extended to the wild life committee and to the advice they offered on conservation.

The Nature Conservancy owed its existence to an unprecedented opportunity ecologists had to both define the problem (conservation of nature reserves and the countryside) and formulate a response by government (a research and management agency led by ecologists). This opportunity was the product of several circumstances. Before World War II neither government nor universities were willing to provide ecologists with the support they considered necessary. In the early 1940s, however, a new possible source for support emerged in the nature reserves movement and government interest in land-use planning. Ecologists argued that nature reserves, and effective land use and management generally, required a commitment to nature conservation and a new institution for ecological research. Convinced by general ideological concerns and by their own experience of the need for scientific autonomy, ecologists, led by Tansley, insisted that this institution be independent. In obtaining government support for their research, ecologists were aided by the postwar prestige of science and the interest in scientific solutions to problems facing British society. Thus, matters of politics, economics, and ideology were all instrumental in the ecologists' campaign for institutional security and in the shaping of the eventual Nature Conservancy.

Interwoven in these events were ecologists' conceptions of their own identity: as scientists with a distinctive perspective on the natural environment, combining experimentation and survey in ways unique to ecology. This perspective was itself guided by a view of nature as made up of complex communities of plants and animals interacting with one another and their physical environment—that is, as ecosystems. By the early 1940s several ecologists had realized that ecosystem study, and new forms of ecological research generally, required both protected areas of land and an institution unencumbered by demands for immediate practical relevance. This realization helped motivate ecologists in their advocacy of nature conservation and freedom in science. In effect, ambitions for their discipline shaped ecologists' efforts to obtain institutional support as it shaped their plans for the conservancy. As I discuss in the following chapter, their success in obtaining this support and making the conservancy a reality would help shape postwar British ecology. And as this comparative history develops it will also become apparent how these circumstances and their outcome—in the form of an independent conservancy responsible for management and research, led by ecologists—differed from the experience of American and Canadian ecologists.

3

Woodland Ecosystems and Nature Reserves

With official acceptance of their arguments for a Nature Conservancy, ecologists shifted their attention from Whitehall back to the countryside. An array of areas of scientific interest lay before them—from the expanses of blanket peat at Moor House in the Pennines of northern England to the enclosed stands of conifers and hardwoods of the Forest of Dean near Wales to the mixed fields of hedges and heathland in Dorset. All these sites and many elsewhere became the subject of study and conservation. At several centers conservancy ecologists studied the soil, flora, and fauna, and addressed problems concerning the management of nature reserves and other aspects of conservation. And as opportunities arose, ecologists advised landowners, industry, and other government agencies on the protection or enhancement of natural features in the British rural landscape.

During the 1950s and early 1960s—the period considered in this chapter—ecologists themselves set the conservancy's research and conservation agendas. They could do so both because of the independent status of the conservancy (a status it sought to maintain during the 1950s) and because of the relative absence of political controversy and conflict concerning nature conservation. As noted in the previous chapter, the insulation of conservation from the national parks debate of the 1940s had helped make it possible for the conservancy, unlike the National Parks Commission, to be equipped with powers of land acquisition and management.

Once established, the conservancy continued to see avoidance of controversy or conflict as a source of its effectiveness. After the war, such major uses of the countryside as agriculture and forestry were usually perceived as consistent with preservation of the traditional features of the British countryside, including wild life and natural areas. This consensus would break down under the pressure of more intensive, mechanized agriculture, as well as urban, recreational, and industrial development. But while it persisted, apparent conflicts between economic activities and conservation were considered to stem not from conflicts in basic values or objectives but inadequate knowledge. With appropriate expertise, land uses could be modified to serve the interests of food and timber production, and of conservation. Thus, the conservancy's mission of providing expert advice to land users reflected the conviction that conservation could be furthered most through greater awareness of ecologically sound land uses.

The conservancy's dual emphasis—on research directed by ecologists and on providing advice concerning conservation—was evident in activities at the largest of the conservancy's first set of research centers, the Merlewood station in the English Highlands. There, ecologists led by J. Derek Ovington pursued a woodlands research program that emphasized plant-soil relations, productivity, cooperation among ecologists, and field experiments, thereby advancing in directions mapped out by senior British ecologists, including Tansley and William Pearsall. The research also opened up new avenues, particularly by drawing on American work in ecosystem ecology. While such research was presented as relevant to Highlands conservation, of more immediate relevance to practical issues were studies, also pursued at Merlewood, on nature reserves management. Studies of woodlands regeneration illustrated how ecological research was linked to specific local conservation issues.

THE CONSERVANCY'S FIRST STEPS

Hampered by a lack of staff knowledgeable about conservation, and perhaps by organizational problems, the conservancy had an uncertain beginning. Out of its first grant-in-aid for 1949–50 of £100,000 it spent only £36,815; not until its fourth year did the conservancy manage to use its entire budget (see fig. 2).[1] At this point finances replaced lack of qualified personnel as the major constraint on the conservancy: during its first decade it received only a frac-

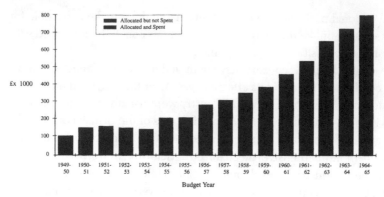

Fig. 2. Nature Conservancy Annual Budgets, 1949 to 1965
Source: *Nature Conservancy annual reports*

tion of the funds allocated to the other research councils. These limits reflected in part the low priority of conservation within the government. But all civil science was restricted. Although science spending rose from £6.5 million in 1945 to £30 million in 1950, this rapid rise ended just as the conservancy began, when government priorities shifted toward rearmament.[2]

The conservancy began to perform more effectively in 1952, when Max Nicholson became its second director-general, replacing Cyril Diver, an accomplished amateur naturalist. By the 1950s Nicholson was an experienced civil servant. During the war he had worked in the Ministry of War Transport, where he was responsible for allocation of shipping. Immediately after the war Lord President Herbert Morrison appointed him secretary of his office, a position he held until 1952.[3] His expertise in field study and its organization (gained through his ornithological work) and in administration proved ideal to the political and scientific challenges facing the conservancy.

One major political challenge was protecting the conservancy's independence. Like the other research councils, the conservancy had considerable autonomy. Although the Committee of the Privy Council for Nature Conservation (chaired by the Lord President) was entitled to direct the conservancy, Nicholson noted in 1957 that it had not yet done so. And in most of its decisions, especially those concerning research, the conservancy had relied on advice from the Scientific Policy Committee, composed largely of eminent ecologists.[4] In practice, therefore, ecologists had considerable discretion in determining the conservancy's priorities, just as Tansley and other ecologists had intended.

This discretion was actively defended. In 1958, for example, when a member of a committee suggested that the conservancy come under the direction of the Ministry of Agriculture, Nicholson described this option as "retrograde."[5] Ecological knowledge and not immediate utility should be the primary aim, he said. As the conservancy noted in 1955, "[Practical applications] ... must be regarded as windfalls or bonuses, and should not be allowed to distort the programme of scientific research."[6] The conservancy nevertheless assured the government that it was aware of practical concerns: when a review committee commented on the "surprisingly large element" of fundamental research in its programs, the conservancy noted that since such research provided the basis for applied research, much of it was therefore itself "applied." The conservancy also stressed its reputation among scientists: this same committee noted that the "consensus of the scientific opinion which we have obtained is, very positively, that the research work being done by the conservancy is valuable, its programme reasonable, and its establishment extremely modest."[7]

The successful defense during the 1950s of the conservancy's independence demonstrated the conviction that scientific activities should be directed by a limited number of senior scientists. Participation of agencies outside the conservancy could "distort" research that should remain focused on long-term objectives. Nevertheless, as I discuss below, the conservancy maintained ties with other agencies, consulting, for example, with the Forestry Commission and the Ministry of Agriculture concerning research policy. Overall, therefore, the conservancy's scientific priorities were set by a small group of scientists who consulted when necessary with scientists and users of scientific information in other agencies.

The conservancy's approach to setting conservation priorities paralleled that established for science. In the 1940s conservation had an accepted meaning in the United States, where resource management professions had adopted the term. In Britain the term signified a generally scientific approach to wild life protection, but no specific program of professional activity yet existed. One principle, however, was clear. Nicholson and other ecologists believed that conservation could be advanced not by confronting land users when their practices conflicted with the goals of conservation but by cooperating with them, explaining how, through the application of ecological knowledge, their activities could advance both their own goals and those of conservation.

This belief epitomized a distinctive feature of conservancy practices and of environmental regulation in Britain generally: adherence to a corporatist model of government–private interest relations, eschewing confrontation and emphasizing cooperation and consensus among a limited number of interests.[8] This strategy was evident in the conservancy's emphasis on cooperation and consensus fostered through application of ecological expertise; in its support of new voluntary organizations that could further its objectives at the local level while demonstrating public support for conservation; and in its inclusion of representatives from agriculture, forestry, political parties, and other interests, thereby allowing potential conflicts between conservation and other land uses to be resolved within a limited forum.

Beyond defining this general perspective the conservancy also had to identify specific priorities. As I have noted, the Wild Life Conservation Special Committee, in its 1947 report, recommended that the proposed nature conservation agency be composed of a "Biological Service" that would conduct research applied to practical problems of conservation and a network of terrestrial research institutes for basic ecological research. The Biological Service never was created, however, perhaps because of the conservancy's initial financial and organizational constraints. Nevertheless, conservation activities got quickly under way. An immediate task was to resurvey potential nature reserves and the smaller Sites of Special Scientific Interest (SSSIs) to update information gathered by the Nature Reserves Investigation Committee and the Wild Life Conservation Special Committee during and immediately after the war. By September 1953 local planning authorities had been notified of 1,098 SSSIs in England alone.[9] Surveys of existing and proposed SSSIs, assessment of the impact of proposed land use changes on them, and advice to owners on maintaining their land's natural or scientific interest became significant foci of the conservancy's activity.

A variety of other conservation projects were also soon under way. Most important, the conservancy moved toward establishing a formal conservation agency that resembled the proposed Biological Service. In 1956 it established a Conservation Branch that was responsible for reserve management and other tasks; it was directed by E. Barton Worthington. Worthington already had wide experience in both basic research and conservation as director of the Freshwater Biological Association and as a researcher and adviser to colonial governments in Africa.[10] By appointing such a senior scientist Nicholson signaled that the conservancy was prepared to give a high priority to conservation. Besides improving reserve management and other aspects of conservation, the Conservation Branch would also ensure that the existing network of scientists would no longer be directly responsible for such tasks, thereby reducing the "conflicting responsibilities of research scientists in applied scientific fields."[11]

In its early years, there had been within the conservancy heated disputes over conservation in general and reserve management specifically.[12] Some naturalists remained convinced that nature reserves did not need science but protection, and they were skeptical of plans for reserve management and experimentation. Influential ecologists, however, successfully argued the importance of active management of reserves. The conservancy had also been advised to give more attention to conservation, including reserve management, when the Treasury inspected the conservancy's administration in 1955. The Treasury suggested that conservation—the "central executive task of the Conservancy"—be the job of people other than those responsible for pure research.[13] The following year a government committee criticized the conservancy for acquiring more reserves than it could effectively manage.[14]

To address these concerns Worthington and his colleagues quickly built up the Conservation Branch. Between 1954 and 1964 the number of conservation researchers and wardens rose from sixteen to eighty-seven. Many became "regional officers," each responsible for conservation within a specific area. For several years a chief concern of regional officers was reserve management plans. Each plan required detailed descriptions of the flora, fauna, and other features of the reserve, its natural and human history, and strategies for maintaining or enhancing its desirable features. After seven years plans had been designed for 70 of the 111 reserves.[15]

Research was developed as quickly as funds and availability of qualified workers permitted. Studies of peat moors, salt marshes, statistical methods, and other topics began almost immediately. The conservancy also received requests from other government agencies to examine specific problems, and by 1952 the conservancy was considering the effects of military activities on sand dunes. That same year the conservancy began studying the effects of pesticides on wild life, initiating what would become, as John Sheail has chronicled, one of its most prominent research efforts.[16]

The centerpiece of the conservancy's scientific plans were its research stations. By 1954 it had established four stations, including Merlewood, in the Lake District. Another was at Furzebrook, near England's southern coast. There were smaller field stations in Beinn Eighe Nature Reserve in Northern Scotland and at Moor House Nature Reserve in the Pennines. The rapid establishment of these stations and of such reserves as Yarner Wood and Moor House that would be useful for experimental studies demonstrated the priority accorded research. Activities at Merlewood, the largest of these stations, provide insights into the conservancy's overall research program.

RESEARCH AT MERLEWOOD

The Highlands of England extend across the country's north, from the Lake District in the west to Northumberland in the east, and down its center almost to the Midlands. Tempered by rugged topography and a rigorous climate, its flora and fauna have attracted naturalists and ecologists for decades, and the presence of five of England's seven national parks testifies to the region's spectacular beauty.

The Wild Life Conservation Special Committee had originally identified four sites in England and Wales for terrestrial research stations. One, to be devoted to the study of alpine, moorland, and other upland communities, would be located in either the Lake District or Snowdonia.[17] Given ecologists' interest in the Highlands, this station became the highest priority in the conservancy's effort to establish new stations, and in 1952 a large house known as Merlewood was bought near Grange-over-Sands, in the south of the Lake District. Renovations delayed its first full season of research until spring 1954. The Moor House station also became associated with the Merlewood station, forty kilometers to the southwest, with its researchers using the larger facility's laboratories. To cement this relationship, Verona Conway, the conservancy's senior botanist and a researcher on peat at Moor House, in 1955 became director of the Merlewood station.

Merlewood's status within the conservancy likely was the result, at least in part, of the influence of William Pearsall. As a boy Pearsall had dredged for plants in the lakes and hiked the hills of the Lake District with his father, a noted naturalist. As an adult, he continued his study of this district, the Pennines, and the North Yorkshire moors, developing an unequaled understanding of the English Highlands. He also introduced several generations of students to its ecology as a professor at the University of Leeds (1919–38), University of Sheffield (1938–44), and University College London (1944–57), as well as through his book *Mountains and Moorlands* (1950). In it he points out both the opportunities the region presents for ecologists and the need for careful, informed conservation.[18] Through the work of Pearsall and others the Highlands were of great interest to British ecologists.

Ovington, a forest ecologist, was assigned to develop the Merlewood woodlands research program (see fig. 3). One of the first ecologists to be hired by the conservancy, Ovington was also one of its most productive researchers. He had received a doctorate in 1947 from the University of Sheffield, where he studied ground flora and leaf litter in woodlands.[19] He then spent two years at the Macaulay Institute for Soil Research, in Aberdeen, studying, with assistance from the Forestry Commission, the effects of afforestation on the ground flora and soil at two commission plantations on the coast of Scotland. These studies were especially noteworthy for Ovington's comprehensive approach. His examination of the soil, for example, considered its moisture, physical and chemical characteristics, and organic and nutrient content, and

Fig. 3. J. Derek Ovington (courtesy of Hunt Institute for Botanical Documentation, Carnegie-Mellon University, Pittsburgh)

he applied the sophisticated methods of soil chemical analysis available at the Macaulay Institute.[20]

Ovington was hired by the conservancy in 1950. Although his first major project was more ambitious than his Scottish studies, its focus was similar: the impact of new forests on soil and flora. What characteristics of plant communities, he asked, were caused by the soil, and how did plants, both individually and within their communities, change the nature of the soil?

In some ways this topic was familiar. The interaction between plants and soil had interested British ecologists since the vegetation surveys of the turn of the century. Tansley himself had considered the problem. In 1935 he first wrote of the ecosystem concept, in part to further understanding of the relations among the plant community, the fauna dependent on it, and the soil and climate.[21] As he explained later, "The study of the habitat should not in practice be separated from the study of the community. The wide field included ought to be envisaged as a whole."[22]

Pearsall had also been attracted to this problem. For more than thirty years he had studied the role of soils in determining the nature of plant communities and the role of plants in modifying the soil.[23] He emphasized the chemical properties of soil—particularly pH as a general indicator of soil characteristics—and demonstrated the role of bacteria and fungi in modifying soil chemistry, thereby linking soil microflora and microfauna research to the study of plant communities. He also studied the cycling of organic matter in woodlands, including the breakdown of leaves and its effect on the soil.[24] It was to this evolving understanding of the relation between plants and soil that Ovington planned to contribute. Such research would serve the scientific goals of British ecologists expressed as the conservancy was being formulated.

For his first three years in the conservancy Ovington was based in facilities provided by Pearsall at University College London. He also spent a great deal of time in the field, particularly at three forests in southern England: Thetford Chase, Forest of Dean, and Bedgebury. At each forest, various species of trees had been planted on adjacent plots at least twenty years previously by the Forestry Commission so that their growth could be compared. Ovington believed that by assuming that soil and climate were roughly the same at all adjacent plots he could compare the long-term impact of different tree species on the soil and flora, particularly since the background of each plot—when it was planted and its climatic and management history—was recorded.

Between 1951 and 1953 he measured the quantity and chemical composition of the mineral soil and litter, the ground flora, and the leaves, boles, and canopies of the trees. He also measured the penetration of rainfall and sunlight through the canopy in the plots at Bedgebury. His results appeared in a series of papers published during the 1950s. The various species of trees, he found, had a pervasive influence on other aspects of the woodlands ecosystem. Tree species largely determined the chemical content and pH of forest floor organic matter, through litter fall and through the influence of the underlying soil. Trees increased soil nutrients, possibly by transporting them through their roots from deeper soil. Composition and weight of ground vegetation was also affected by the distinctive microclimates under the canopies of each tree species and by changes in soil characteristics. Beyond these specific conclusions, Ovington's research confirmed to his own satisfaction that the components of woodlands interacted in ways that could be fully understood only through integrated ecosystem study.

At Merlewood, Ovington continued to analyze these results, focusing on the complex interaction between vegetation and soil. He studied how leaves accumulate on the forest floor and then decompose to form new soil, completing the cycle by which nutrients taken up by trees are returned to the soil to provide for future growth. Ovington's measurements suggested that this process did not occur at the same rate in all forests: the deeper accumulation of litter under conifers indicated that it decomposed more slowly than did hardwood litter.[25]

To explain these different rates it was necessary to study the role of soil or-

ganisms in litter breakdown and decomposition. This required the cooperation of ecologists with expertise in zoology, botany, and soil science. At Merlewood several ecologists combined their efforts: O. J. W. Gilbert analyzed the
nutritional value to soil fauna of various species of fallen leaves, J. E. Satchell
studied the distribution of micro-arthropods in different soil types, C. K. Capstick examined soil nematodes, J. S. Waid developed methods of measuring
the abundance and activity of soil fungi and other forms of soil microflora, and
M. S. Laverack studied the contribution of earthworms to decomposition.[26]
By relating rates of litter decomposition to specific soil flora and fauna they
identified the soil communities characteristic of particular woodland types.

Such collaboration became possible only after the conservancy was created, fulfilling the vision of Tansley and other ecologists, who had argued that
collaborative group research was necessary for effective study of ecosystems.
Such collaboration was often impossible in British universities. As Max
Nicholson noted in 1958, "There is an element in the organisation and structure of the universities which makes it very difficult for them to co-operate ex-
tensively. . . . Most of this work involves both botany and zoology, and often
soils too, and those are the provinces of different departments in the university, each having their own . . . research programme; and these synthesising
subjects, which involve co-operation in a team between members of different
groups, are very difficult to organise in the university."[27]

Ovington and his coworkers also began several new projects, including a
survey of the nearby Roudsea Wood Nature Reserve and experimental applications of nutrient mixtures on small plots to simulate the effect of leaf-falls on
soil and ground flora.[28] Another project was a long-term study in the Forestry
Commission's Gisburn Forest. This comprised a series of experimental plots,
some planted with various species of hardwood and coniferous trees, some
left unplanted, and others set aside for sheep grazing. Its purpose was to
measure plant productivity and changes in soil conditions as the trees matured. By tracing the changing productivity of trees Ovington demonstrated
one method of integrated study of the effects of physiological processes and
environmental factors (including changes in soil) on individual trees as they
grow from seedlings to maturity.

The Gisburn Forest study and Ovington's use of forestry commission plots
illustrate how Merlewood research reflected the ambitions of those ecologists
who had advocated establishment of the conservancy. Such long-term experiments helped push British ecology away from descriptive survey and interpretation of natural vegetation communities and toward manipulation of
seminatural or artificial communities. The merits of experimentation became
a matter of debate among ecologists. Some were wary, arguing that the results

could be unreliable because "stands of single tree species give rise to such artificial ecological conditions that production in plant and animal populations may not be in balance."[29] But others welcomed it. George Varley, for example, used his presidential address to the British Ecological Society to criticize ecologists for ignoring the large-scale "experiments" being undertaken by agriculturists and foresters in favor of natural habitats less obviously changed by human activity. Ecologists, he argued, should make use of such experiments.[30]

The Gisburn Forest plot experiments were intended for the long term (they continue to the present day). To provide more immediate results Ovington compared the growth of Scots pine plantations at Thetford Chase and natural forests of silver birch at Holme Fen Nature Reserve, near Peterborough. By cutting down trees of varying ages and weighing their parts he measured the changing rate of production of leaves, branches, and bole as individual trees matured. In comparing the two forests he concluded that because the natural forest is more dense (with spacing controlled by competition between trees) than the artificial plantation, it would tend to reach maximum productivity more rapidly. The eventual productivity of a conifer forest, however, would be greater because those trees carry leaves throughout the year and can grow whenever conditions are favorable.[31]

This comparison of conifer and hardwood forests was part of a new research program on woodland productivity that Ovington developed as he completed his study of how trees influence soil. Like many other British ecologists he became interested in the 1950s in "production ecology": the study of factors influencing the growth and productivity of plant and animal communities. Ecologists studied, for example, the respiration of soil fauna to estimate their energy consumption; the production of birds and their prey; plankton and fish production; and the productivity of grasslands. Among plant ecologists there was particular interest in the processes and products of photosynthesis and in relating plant productivity to the efficiency of their capture and use of light.[32]

As he had been in the study of Highland and soil ecology, Pearsall was a leading proponent of production ecology. He saw it as a means of maintaining an integrative perspective for ecology. Considering the shadow cast by the towering prestige of cellular and molecular biology, he argued in 1954 that botanists must struggle to integrate their results, however "scientifically unfashionable" the effort would be. This, he explained, would balance the specialization engendered by new techniques such as electron microscopy and microchemical analysis: "We often do not know what our results mean, and it will doubtless become more and more important to stand back from time to time and ask ourselves where we are going and also how we can integrate the results we are obtaining."[33] Ecologists could maintain their distinctive, nonreductionist identity through study of the growth of entire organisms within their environment.

Throughout the 1950s Pearsall's influence at Merlewood was readily apparent. Much of the research there and at Moor House on plant-soil interactions and productivity extended inquiries he had pioneered. This is not surprising, as Pearsall was a "constant visitor" to Merlewood and to Moor House throughout the 1950s and was also (from 1954 to 1963) chairman of the conservancy's Scientific Policy Committee.[34]

In the late 1950s Ovington increasingly invoked the ecosystem concept in his efforts to develop an integrative perspective on woodlands. His interpretation of the concept, however, marked a departure from Tansley's view. For Ovington it meant understanding woodlands in terms of the cycling of nutrients. Nutrients are lost from the ecosystem through tree removal and leaching; they are gained through mineral weathering, precipitation, and dust. Their movement within the ecosystem—uptake by vegetation, fall to the forest floor in leaves and other debris, and return to the soil—was also crucial. This view of ecosystems in terms of nutrient cycling had two functions. It ordered the highly complex interactions between plants and soil that Ovington (and other ecologists) had studied since the late 1940s, and it showed where research was needed in order to identify how nutrients moved within the ecosystem and at what rates. This was the chief purpose, for example, of studies of soil fauna: elucidation of its role in the movement of nutrients from litter to the soil.[35]

Ovington looked to the United States for ideas concerning ecosystem ecology, particularly to Eugene Odum's textbook *Fundamentals of Ecology* (1953), studies of productivity by George Riley, G. Evelyn Hutchinson, and others, and research on forests and hydrology by foresters.[36] He also crossed the Atlantic to learn first-hand about American research. In September 1956 he visited experimental watersheds, where researchers were studying the role of forests in regulating waterflow.[37] And two years later he took a leave of absence to become acting director of the Terrestrial Ecosystem Project of the University of Minnesota's department of botany.

In Minnesota, Ovington applied another approach being developed by American ecosystem ecologists: the measurement of energy flow. By measuring the energy content of samples of plant organic matter using a bomb calorimeter and then extrapolating these data to the total production of a given land area, he could determine the flow of energy through an entire woodland. He then compared this to the incident solar radiation in the forest to estimate the woodland's photosynthetic efficiency. Using samples from the forest plots he had studied in the early 1950s and the pine plantations at Thetford Chase he estimated the efficiency of coniferous woodlands at about 2.5 percent. That is, about one part in forty of the incident solar energy was converted to organic matter.[38]

When Ovington returned to Britain in 1959 he went to the conservancy's London headquarters to become a principal scientific officer, responsible for woodlands. He continued to develop his ideas concerning integrative, quantitative studies that could transcend individual organisms and species to consider the fundamental processes of woodland ecosystems. As he explained in 1962, "If forest research is to be fully effective, it needs to be orientated towards obtaining a better appreciation of ecosystem dynamics; particularly in relation to quantitative studies of the biological and physical processes affecting productivity and the accumulation, transformation and flow of energy and materials (water, mineral elements, etc.) through different woodland ecosystems. . . . These processes constitute an expression of the woodland ecosystem perhaps more significant than the trees themselves."[39]

Reviewing his own and other researchers' experiences, he identified four aspects of woodland ecosystems requiring study: organic matter production, energy flow, circulation of water, and circulation of chemical elements. While often considered separately, each, Ovington asserted (using language characteristic of Eugene Odum and other American ecosystem ecologists), was really only an aspect of the "fundamental unity of ecosystem physiology," and it was through the ecosystem approach that the "most challenging and rewarding problem facing woodland ecologists"—synthesizing fragments of knowledge to form a comprehensive understanding of woodlands—could be accomplished. After twelve years of research within the conservancy he could not alone provide such a comprehensive understanding. Nor, he predicted, was one likely to appear soon. But he was confident that by combining the insights of different disciplines, by viewing woodlands as dynamic ecosystems amenable to quantitative study of the movement of matter, energy, and water, as had been done at Merlewood, ecologists could best understand their "patterned complexity."[40]

In the early 1960s Ovington moved to the conservancy's new Monks Wood Experimental Station to direct its woodlands research program. He left the conservancy in 1965 to become professor of forestry at the Australian National University in Canberra. His influence continued to be felt, however, in the research program at Merlewood, which in the 1960s still emphasized plant-soil relations and woodlands nutrient cycling, and in the European component of the International Biological Program (IBP), which, in its focus on production ecology, continued research in areas he had promoted since the early 1950s.[41] Ovington was active in forming the IBP, in 1964 cowriting with J. B. Cragg a proposal for a terrestrial production group within the program.[42]

Several elements of Ovington's conservancy research are noteworthy. It exhibited a gradual shift in focus from organisms within their physical environment (as reflected in his studies of plant-soil relations) to ecosystems, in which processes that were "perhaps more significant than the trees themselves" became the unit of study. This shift was in effect a move away from Tansley's concept of ecosystems, in which individual organisms retained a strong identity, and toward the emerging perspective, developing especially in the United States, of ecosystems in terms of functional components and processes, not organisms or species. I consider the development of this ecosystem perspective in greater detail in subsequent chapters.

Frank Golley has suggested that the development of ecosystem ecology has been "largely an American tale."[43] However, while Ovington's ecosystem perspective reflected his contacts with American ecologists, it was also shaped by ideas from closer to home, including studies of soil chemistry and production ecology. His work, and the work of the Merlewood station generally, demonstrate the significance of ecosystem ecology in British ecological research of the 1950s and the distinctive character of this work.

CONSERVATION AT MERLEWOOD

In the south of the Lake District, at Morecambe Bay, lies Roudsea Wood National Nature Reserve. Within its sixty-nine hectares lie a range of soil conditions—from one ridge of acidic slate down to a damp valley floor and up again to another ridge of limestone—supporting many species of ground flora and woodlands dominated by oak, ash, and birch. These communities have attracted much attention from naturalists, who have urged their protection since at least 1915.[44] It was also the first woodland reserve to receive a comprehensive management plan. While its natural features made it a priority, a plan was also needed to ensure its effective use for research, because it is only about ten kilometers from Merlewood. Preparation of this plan, accordingly, was one of Merlewood's first conservation responsibilities.

Research in Roudsea Wood began promptly after Merlewood opened, with surveys of its plants, soil, and topography.[45] Its management plan, approved in July 1957, set as a long-term goal the establishment of a native ash–oakwood forest characteristic of north-west England. Such a forest would be very different from what then existed in the reserve. Most of the forest consisted of coppice, a traditional form of woodlands management in which trees are cut off close to the ground and then permitted to grow again out of the stumps. Typically, each tree of a well-established coppice wood will have many smaller trunks emerging out of a single massive stump.[46] Coaxing a coppice wood toward a native woodland required a great deal of intervention, including the gradual cutting off of all but one of the stems on each stump, with the one remaining being allowed to grow to maturity. Ultimately there would be an "entirely self-sown and self-perpetuating wood of varied canopy height and age of composition."[47]

There was little experience available to aid in designing such a strategy. As a result, management practices at Roudsea Wood, as at other reserves, were often designed as experiments in which a variety of methods would be tried to discover which was best. Under the direction of R. J. Elliott, the regional officer based at Merlewood, some areas of the woodland were not managed at all and one was maintained as coppice, which could be compared with the unmanaged and managed native forest areas.[48]

One immediate problem encountered was that few natural seedlings were surviving. Because natural forests depend on self-regeneration, this problem stimulated much research. Studies included surveys of acorn fall, germination, and seedling survival, as well as experiments on the effects of various factors on the growth of oak seedlings. One study focused on caterpillars believed to feed on oak leaves and flowers. Another used protective cages to establish that rodents and birds were responsible for seed loss and damage to seedlings.[49]

Merlewood ecologists were heavily involved in this research. As Worthington and other conservancy ecologists had intended, there was considerable contact between researchers and the regional officer and warden responsible for reserve management. Although the researchers were not part of the conservation branch of the conservancy, many of its regional officers were based at research stations and held job classifications permitting interchangeability between research and conservation staff. This was to ensure that researchers "share the stimulus and responsibility of solving actual land management problems."[50]

The problems encountered in managing Roudsea Wood, and the research addressing these problems, reflected particular conditions in this region of Britain. Reserve management and research generally were tied tightly to the local context of nearby ecological and human communities. Management practices—mowing grasslands, burning or cutting peat, coppicing woodlands—reflected both local ecological conditions and past practices. Reflecting the richness and complexity of the natural and human history of the British countryside, specific practices were often, of necessity, appropriate only to a particular region. Accordingly, conservancy officers had to be based in each region because conservation depended on understanding both the ecology of the countryside and the local customs. As Nicholson explained to a parliamentary committee, "Matters to do with land are rather intimate and local, and there is an astonishing variety of land tenures and outlooks and practices in relation to land, and we always think it best to deal with the landowners and the occupiers and the other people interested in land in each region by people who are themselves located in that region. For instance, our officers in Wales speak Welsh; our officers in the North Country speak with a North Country accent. In Scotland our officers wear kilts."[51]

The need for conservation research and advice to be sensitive to local conditions was emphasized by the conservancy's reliance on county trusts: local landholding voluntary organizations that with professional guidance, and for little extra expense, extended the conservancy's reach. By 1965 every part of England and Wales was covered by a county trust, each responding to specific local conditions and priorities.[52] Its relations with these local agencies epitomized the conservancy's emphasis on cooperation in furthering its conservation objectives.

In contrast, Ovington's research on plant-soil relations, production ecology, or ecosystem ecology was not immediately relevant to reserve management. He was aware of this: when reviewing woodlands management practices, he emphasized practical strategies developed or refined by management-oriented researchers.[53] Production ecology, however, was relevant to the conservation of marginal land; even if unsuitable for agriculture, such land could, with appropriate vegetation, be highly productive. Study of plant-soil relations was also valuable, as it provided a basis for maintenance of woodland soil fertility.[54]

This view echoed that of Pearsall's concerning the English Highlands. He had identified two key problems in their conservation: enhancement of their productivity and maintenance of their native flora and fauna. Solving these problems, he argued in 1953, required an understanding of the "natural trends of soil and vegetation," including, for example, soil changes resulting from deforestation or afforestation. Because loss of trees could lead to loss of soil fertility, perhaps the best way to restore fertility and raise the productivity of these marginal areas would be afforestation.[55] In *Mountains and Moorlands* he had argued that forestry policy should aim for the restoration of the natural fertility of moorland soils through fertilization and introduction of deciduous species. More generally, as Pearsall explained elsewhere, proper land use required identifying which type of vegetation—coniferous woodland, grassland, or swamp plants—would be most productive in a given area.[56]

Ovington especially stressed the relevance of his work to forest management. As I have noted, he did much of his research in Forestry Commission forests. These were part of the commission's long-term plan to develop, over the next fifty years, 3 million acres of new forests and to manage another 2 million acres of existing forests.[57] These forests would be very different from natural forests, being composed of a single species and planted in uniform, even-aged blocks. Usually they would be fast-growing conifers instead of the deciduous trees characteristic of most British forests.

Ovington suggested that his research could contribute to these plans by identifying, for example, which tree species could best maintain soil fertility.[58] Most of the land available for new forests was relatively infertile, and the impact of such exotic conifers as pine and spruce on that soil was only poorly understood. The commission, for its part, recognized the significance of soil nutrients and was concerned about the effects of exotic conifers.[59]

Even though it was potentially relevant to Forestry Commission concerns, Merlewood research was clearly not guided primarily by the commission's needs. It dealt with only one issue of interest to foresters. They recognized the risk of soil impoverishment and the need for understanding the relation of trees to the soil, but foresters were also concerned about such problems as fire, wind, disease, and the soundness of the timber.[60] Another concern was the effect of rabbits, squirrels, insects, and fungal and bacterial diseases on forest productivity. According to some assessments, such pests were a more immediate and serious problem than were changes in the soil. In general the Forestry Commission favored short-term research directed toward problems encountered in planting new forests rather than toward the problems of maintaining established plantations.[61]

Merlewood ecologists considered their research relevant to a variety of environmental problems of the British countryside. Some studies addressed immediate problems encountered in reserve management; others, particularly Ovington's, were considered relevant to more general issues of appropriate land use. Institutional links between research and management appear to have helped shape this pattern: while researchers worked beside reserve managers and responded to their immediate concerns, the studies of plant-soil relations, production, or ecosystem ecology that conservancy ecologists considered to be relevant to the Forestry Commission actually addressed general, long-term concerns about forest management, not the specific issues that foresters themselves considered most pressing.

RESEARCH AND CONSERVATION IN THE CONSERVANCY

The conservancy's interest in the practical implications of ecology had several sources. It had, of course, a mandate to manage reserves and provide advice to the government. And as Tansley had stressed for many years, ecology had to demonstrate practical significance to ensure continued government support. Success in achieving an independent institution did not eliminate that imperative, especially considering the uncertain support from government for conservation during the 1950s. In addition, as Tansley had noted long before, ecology could benefit from contact with practical concerns, gaining new research problems and methods.

Ecologists themselves wanted to act on their concerns about the British countryside and environmental problems generally. Ovington, for example, contributed to the International Union for the Protection of Nature (after 1956, the International Union for the Conservation of Nature), as a member and then honorary secretary of its Commission on Ecology. His personal concerns no doubt motivated him, as they had Pearsall, Tansley, Nicholson, and others, to develop research relevant to conservation.

These concerns extended to the place of science in national affairs. Nicholson, in particular, was convinced that Britain's development should be based on rational scientific principles. In 1931 he helped prepare "A National Plan for Britain," a precursor to a new organization called Political and Economic Planning (PEP). Over the next two decades PEP produced reports advocating, among other goals, a rational approach to planning Britain's future. Nicholson played a leading role in PEP before and during his tenure at the conservancy, as general secretary from 1933 to 1940 and as a member of its governing body from 1931 to 1978. He also edited many of the reports published by PEP in its initial years. Within PEP, Nicholson helped set up TEC Plan, which promoted the concept of a single national plan and of planning itself as a comprehensive discipline that would consider all issues relevant to economic and land development. He also promoted PEP's 1965 study of land-use planning and countryside development.[62]

Nicholson's activities in the conservancy and in PEP were entirely separate.[63] But they both originated in his commitment to rational decisionmaking based on research and his belief that bad decisions were the product of inadequate information. He had had this view since at least the 1940s, when he had argued to the Wild Life Conservation Special Committee that wild life is damaged most through ignorance. The necessity therefore was not legislation but education and expert advice backed by research.[64] Such convictions, relevant to public policy and to land management, helped drive the conservancy's development of conservation and related research and its application of this research through consultation and cooperation with other agencies.

Beginning at its first meeting in May 1949, when it was consulted about a proposal to introduce a small herd of reindeer into Scotland, the conservancy volunteered advice to and received a succession of requests for comments from various agencies—the Ministries of Agriculture and Defense, other government departments, and county councils and other local authorities—concerning activities likely to affect plant and animal populations or habitats. The conservancy conducted numerous small research projects or collected data to support its advice, which it communicated through interdepartmental meetings, memos, or other means. Throughout, the guiding principle was that scientific information provided a basis for consensus on the conduct of activities by public and private agencies that could satisfy both the agencies' objectives and those of the conservancy.

But while scientists at Merlewood linked their research to conservation requirements, much of their work also aimed for a general understanding of ecosystems. This was especially evident in Ovington's work in plant-soil relations, production ecology, and ecosystem ecology, in which he did not seek solutions to specific local problems of conservation but rather sought basic, general principles of ecosystem structure and function. Although his work, he expected, could help guide land use in marginal environments, practical implications could not be its primary justification. Overall, the opportunities that Ovington enjoyed—the Merlewood facilities, a leave of absence to work in the United States, the latitude to pursue research not directly relevant to reserve management or other immediate tasks—reflected conservancy ecologists' convictions concerning basic science and the need for independence of researchers who had a demonstrated record of research activity.

This pursuit of basic research also reflected the context of British science. The conservancy and the Forestry Commission kept in touch in matters concerning research, establishing in 1954 a Joint Research Liaison Committee on which Ovington was one of two conservancy representatives.[65] Its purpose was to ensure that each organization did not duplicate the other's research. Because of the financial stringency experienced by British civil science in the 1950s, the conservancy risked criticism if it engaged in work that was within the purview of the commission, the Agricultural Research Council (ARC), or other agencies. Duplication was a major concern of an interdepartmental committee reviewing the conservancy in 1956, and there had been "some feeling on the official research side that the Conservancy might be casting its net too wide."[66] The need to avoid duplication was also stressed in the first parliamentary debate on the conservancy, in 1961.[67]

Concern about duplication encouraged conservancy ecologists to emphasize basic research over studies of immediate relevance to agriculture or forestry. For example, because studies of grazing on grassland could overlap the work of the ARC, ecologists restricted their research to fundamental studies of semi-natural types of vegetation.[68] Overall, as the conservancy could not duplicate the applied research of the Forestry Commission or the ARC, this left basic ecological research as the area that the conservancy could call its own. Paradoxically, therefore, the conservancy's links with such agencies as the Forestry Commission that might have been expected to stress applied research needs actually encouraged the conservancy to emphasize basic research, except in areas (such as nature reserves) in which the conservancy itself had practical responsibilities.

It was nevertheless asserted that basic research could contribute to forestry, farming, and other land uses. A "fundamental understanding of how nature works" gained through ecological research, Nicholson explained, could be as useful as the "important contributions in the applied field stimulated by the Agricultural Research Council." Overall, the need to demonstrate the utility of basic ecological research, while not duplicating the work of the ARC or the Forestry Commission, led to a broader definition of conservation. In 1949, conservation other than reserve management remained a field "largely unexplored." In the 1950s the conservancy extended its scope beyond reserve management to encompass provision of a "sound and well-balanced technical basis for decisions about land use in this crowded country." Agricultural land was usually excluded from this definition: "Prima facie, land which is agricultural land is land in which the Nature Conservancy is not interested and therefore the field of possible conflict between nature conservation and agriculture is very limited."[69] Conservation was, however, relevant to the less economically valuable land in Britain: moorlands, woodlands, grasslands, heaths, and coastal areas. Many scientists, including Pearsall, Worthington, and other ecologists, noted the need to apply scientific principles to prevent misuse of this land.[70] In particular, Merlewood research and studies at Moor House Nature Reserve could contribute to conservation of the British Highlands, where crop production was difficult or impossible.

Choices made by conservancy ecologists in the 1950s therefore reflected two related goals: to develop basic ecological research and to build the applied science of conservation. One link between these was a concern for the long term. Research projects sometimes continued for years, even decades. Ovington's Gisburn Forest experiments, for example, have continued for four decades. Other projects on moorland ecology and on the ecology of grouse also continued for long periods.[71] Similarly, conservation sought long-term solutions, whether in maintaining the productivity of marginal land or in managing reserves. If others based their decisions on the benefits to be obtained today, then it fell to ecologists like Ovington to consider tomorrow and the day after tomorrow: "If, over centuries, a steady crop of timber is to be taken from forested lands, then the average outgo of nutrients must not exceed the average income if soil impoverishment is to be avoided. In view of the present widely spread economic forestry policy of growing intensive plantations of conifers on a short-term basis—largely for pulping, etc.—this may become a matter of considerable importance."[72]

Management plans for nature reserves also demanded patience: for example, the conversion at Roudsea Wood of a coppice wood to a high forest would take many decades. Such a project could break down if the managers themselves did not also remain in place. A reserve should therefore be managed by a naturalist who may give a lifetime of study to one reserve: "Continuity of service may become of almost paramount importance."[73]

Long-term concerns reflected ecologists' understanding of the history of the land. The countryside was the product of centuries of change and development: vegetation succession, soil development, and human activity. If reserves were to maintain the best of Britain's flora and fauna, then the long term must be kept in view (particularly in evaluating the success of their management) because the "art of conservation is long and the life of the Conservancy has so far been very short in terms of the slow cycle of natural processes on which Reserve Management Plans must be developed."[74] This concern for the long term, and for historical factors in ecology, also helped justify ecologists' insistence on scientific autonomy. While receptive to requests for information on specific problems, they believed that the conservancy's primary research program at Merlewood and at other research stations should not be shaped to fit the short-term concerns of other agencies.

The relationship between conservancy research and environmental issues in the British countryside was complex, and divergent views were expressed regarding the relationship. On the one hand, conservancy research was responsive to the concerns of British ecologists (and nonecologists who shared these concerns) about the protection and management of natural habitats. As a result, the conservancy generally addressed concerns that could be most readily addressed by ecological expertise. Research conducted at Merlewood demonstrates this. Guided by the Scientific Policy Committee (particularly by Pearsall, who had a special interest in it) and by the requirements of the regional officer stationed there, it contributed to the management of nearby reserves and to the conservation of the economically marginal areas of the

English highlands. This research would be applied through processes exemplifying cooperation and consultation among interested parties, grounded in the idea that consensus concerning conservation and other land uses could be reached through provision of expert ecological information.

Thus these conservation efforts led to what Philip Lowe has described as a "fusion of pure and applied science" that "enabled ecology to break out, much more quickly than other disciplines, from the ivory tower mentality that had characterized academic biology before 1939."[75] Similarly, several conservancy officials, including Nicholson, Worthington, and Norman Moore, have argued that the conservancy's joint responsibility for research and for conservation helped generate conservation ecology, a practical new scientific field.[76] That the conservancy was led by ecologists was crucial, as it helped ensure that the conservation issues selected for attention would be those most relevant to ecological expertise.

At the same time, a considerable amount of Merlewood research, and especially Ovington's studies of woodlands, was of only indirect relevance to conservation. Rather than aid in the solution of problems unique to the region, Ovington sought a general understanding of woodlands through studies of plant-soil relations, production ecology, nutrient cycles, and other ecosystem phenomena. In doing so he contributed to a major goal of British ecologists: the development of ecological research that in its use of experiments and physical and chemical principles adhered to the standards of rigor of the larger scientific community but that employed a distinctive integrative approach to living organisms. While advancing lines of research already mapped out by other British ecologists, he also developed his own approaches, adopting ecosystem perspectives from elsewhere, particularly the United States. His work reflected the latitude that he and a few other conservancy ecologists enjoyed, a freedom that allowed them to take the initiative in developing research directions. This was consistent with John Sheail's conclusion that conservancy research was largely independent of management requirements. Sheail also noted that reserves were often chosen and managed on the basis of their value to ecological research, not for the protection of rare species. Researchers and reserve managers generally worked separately, with researchers pursuing topics of little interest to managers, who often resented the support given research.[77]

Overall, we may interpret the dual imperatives of conservancy research as those of practical relevance and of relevance to the scientific ambitions of ecologists; those imposed by the highly specific conservation problems of particular parts of the British countryside; and those implied by the ambition to build a general science of ecology. The parallel research efforts at Merlewood suggest that these contrasting imperatives may not be readily reconciled within the same research program. Both imperatives, however, were defined in terms that could be most readily addressed by ecologists, thereby enhancing their authority as experts on conservation in the British countryside. This underscored the fact that the conservancy was an institution led by ecologists.

This account ends in the early 1960s, when several trends had begun that would greatly affect ecological research and conservation in Britain. One trend that had an immediate impact was political: the increasing demands for reorganization and redirection of civil science. The goal was to enable research organizations to respond more effectively to demands for scientific expertise. British institutions for conservation and ecology were transformed through the creation in 1965 of the Natural Environment Research Council, with the Nature Conservancy as one component, and the 1973 division of the Nature Conservancy into the Nature Conservancy Council, responsible for conservation, and the Institute for Terrestrial Ecology, whose mandate was ecological research. Most important, the distinctive feature of the Nature Conservancy—its combination of research and conservation within a single institution—was lost.

Changes in the countryside imposed their own strains on British conservation. Whereas in 1949 farming had been viewed as compatible with the traditional, distinctive character of the countryside, by the 1970s the impact of intensive, highly mechanized agriculture had become evident in the elimination of grasslands, hedges, wetlands, and other areas. It also became clear that the conservancy's exclusion of agriculture from its mandate severely limited its influence over what had become the most significant factor damaging wild life habitats in Britain. Finally, the politics of nature conservation and land-use issues in the countryside generally became increasingly marked by confrontation between conflicting values and priorities, the reconciliation of which would often demand more than ecological expertise.

In the next chapter I examine the development of ecology within a very different political, institutional, and scientific context as I discuss an important episode in the history of American ecology.

PART TWO

United States

4

Ecosystems and the Atom at Oak Ridge National Laboratory

When Julian Huxley traveled to the United States in 1955 he encountered in Tennessee an institution that was both a powerful symbol of postwar American leadership in science and unlike anything he had encountered in British ecology and conservation. "On this visit," he recalled, "I was much impressed by the Atomic Energy establishment at Oak Ridge in the Great Smoky Mountains, though my lack of background knowledge of physics interfered with my understanding of the fantastic progress being made there."[1] By that time the Oak Ridge National Laboratory was one of the largest research institutions in the world. Created for the Manhattan Project, it had been the site of a plutonium-producing reactor, uranium isotope separation plants, and the project's administrative headquarters. With peace and the advent of the Atomic Energy Commission it became a center for studies of nuclear energy, basic physics, and radiation biology. It and other AEC national laboratories epitomized the "springtime of Big Physics," when physicists dominated the American research community.[2]

Huxley's difficulty in comprehending the work at Oak Ridge was understandable: British science could not match the scale of its technological and scientific activity. The contrast between it and the Nature Conservancy was especially striking. Oak Ridge had several thousand physical scientists, engineers, and biologists engaged in prestigious technological and scientific research; the conservancy was far smaller, and housed ecologists engaged in much less prominent work. And yet there were similarities between Oak Ridge and the conservancy in their support for ecological research; in their organization and direction of science; and in the political context of scientific expertise at each institution.

Huxley did not mention finding ecologists at Oak Ridge, but they were there. The year before his visit Stanley Auerbach, the first ecologist employed by the laboratory, had begun studying radioactive contamination near waste disposal sites. During the next fifteen years Oak Ridge ecologists expanded from this small beginning, tracing the movement of radionuclides within several habitats. In doing so Auerbach and his colleagues contributed to a range of AEC and Oak Ridge objectives, and support of their research grew steadily. By the end of the 1960s Oak Ridge was home to one of the largest ecological research programs in the United States. There were both abundant facilities for ecological research (radioactive materials, computers, and land) and challenging environmental problems created by the laboratory's concentration of nuclear technology. It resembled the Nature Conservancy in this respect as well, in that both ecological research and environmental problems to which it could be applied were within the same institution, though the problems themselves were very different.

The AEC was one of the largest supporters of ecological research in the United States. Even before the AEC had been established, ecological surveys or research had been initiated at a few Manhattan Project sites.[3] By 1950 the commission was funding research at several universities. In 1955 the Environmental Sciences Division of the Division of Biology and Medicine was created. AEC support for ecological research in commission laboratories and universities continued to grow throughout the 1950s and 1960s, helping to create the new field of radioecology. The commission fostered innovations, such as ecosystem research and the use of systems analysis, radionuclides, and other tools, that helped shape contemporary ecology even as its other activities abetted the emergence of modern environmentalism. Environmental concerns became prominent in American society in part because of fears about test-bomb fallout and nuclear reactors.

The organization and direction of science at Oak Ridge has often reflected changes in federal science policy. Once ecologists became securely established within Oak Ridge they had considerable latitude in establishing their own research objectives. The AEC, like other federal agencies, accorded its researchers both generous funding and autonomy. This attitude reflected a view widely accepted by researchers and policymakers in the two decades after World War II: that scientists, adequately funded and accorded autonomy, would in return provide ideas and products of benefit to the nation. The similarity between this view and the British government's willingness to accord the conservancy its autonomy is readily evident. But there was an important difference: in the America of the 1950s the invocation of scientific autonomy obscured the focus of postwar science on the economic and social priorities of the national security state.[4] At Oak Ridge the AEC's own ideal of a new technological and organizational basis for society, fueled by nuclear energy and contributing to the nation's economic and military security, shaped the research choices available to ecologists. And Oak Ridge ecosystem ecology emerged at a time when belief was strong that application of expert knowledge, backed by government resources, could ensure rational management of society and its environment. Consistent with this view, Oak Ridge ecologists redefined their objects of study, viewing nature as an assemblage of ecosystems susceptible to comprehensive analysis and management.

The autonomy of research at Oak Ridge in the 1950s and 1960s also reflected the political context of nuclear technology. The AEC, its congressional overseers on the Joint Committee on Atomic Energy, and the nuclear industry epitomized a characteristic tripartite pattern of American government: "iron triangles," composed of a federal agency, a legislative committee, and an interest group, that share at least a rough consensus concerning their objectives within a circumscribed policy area shape activities intended to further these objectives and tend to exclude those outside the triangle.[5] Within an iron triangle expertise will tend to be insulated from the pressures of a pluralistic political framework. In the case of the AEC, the insulation of nuclear expertise from political controversy was analogous to that of ecological expertise within the conservancy, which was applied in the form of advice to other agencies and voluntary groups and insulated from controversies concerning countryside planning or the ecological impacts of agriculture.

A FOOTHOLD FOR ECOLOGY

In 1942 thousands of military and civilian workers invaded a remote area of Tennessee to build a plutonium-producing reactor and uranium isotope separation plant for the Manhattan Project. These facilities, the nucleus of the eventual Oak Ridge laboratory, exposed many workers to unprecedented radiation risks. Recognizing the inadequacy of existing methods of radiation detection and shielding, a small group of physicists and physicians began to develop new methods. They coined the term "health physics" to describe their work on radiation detection and design of safeguards.[6]

Karl Morgan, the senior health physicist at Oak Ridge, was among the first to identify a need for ecological research at the laboratory. He defined health physics broadly to include not only study of the immediate dangers of radioactive materials but of the risks posed by contamination of the environment. Because, he wrote, radiation damage could occur "either directly to man or indirectly through the ecology of his environment," both field surveys and studies of the "ecological effects of radiation" were necessary.[7] By 1943 Morgan had begun monitoring radioactivity in waste disposal areas at Oak Ridge.

After the AEC assumed operation of the laboratory Morgan sought to expand the role of ecology within health physics. In 1948 he arranged for Louis Krumholz, a Tennessee Valley Authority fisheries biologist, to survey the distribution of radioactivity in the water bodies and landscape near the laboratory. But his survey did not convince Oak Ridge management that this form of research should continue, and it was terminated in 1953.[8]

The following year, however, another opportunity for ecological research emerged, in the revised Atomic Energy Act of 1954. The act committed the AEC to cooperate with private industry in developing power reactors. Within a week of its passing President Eisenhower had inaugurated construction of the first American nuclear power plant in Shippingport, Pennsylvania. It also, indirectly, encouraged AEC laboratories to seek new missions that industry would be unlikely to undertake.[9] For its part the Oak Ridge Health Physics Division initiated a study of the disposal of wastes produced by power reactors. Edward Struxness, chosen by Morgan to direct this study, shared Morgan's view that health physicists must consider environmental aspects of radioactivity. Struxness had also studied with Orlando Park, an ecologist at Northwestern University. Morgan and Struxness obtained approval to hire an ecologist for the waste disposal program and sought Park's advice. Park suggested Auerbach, a former student who had received his Ph.D. in 1949. Auerbach arrived at Oak Ridge in August 1954 to become an associate scientist in the Health Physics Division (see fig. 4).

The initiation of an ecology program at Oak Ridge indicated the attention the AEC was beginning to pay to the subject in the early 1950s. Ecological surveys or research had been initiated at a few Manhattan Project sites even before the AEC was established.[10] By 1950 the commission was funding research at several universities. In 1955 John Wolfe, a plant ecologist at Ohio State University, joined the AEC's Division of Biology and Medicine. Three years later he became chief of a new Environmental Sciences Division. Under

*Fig. 4. From left, Jerry Olson, Orlando Park, and Stanley Auerbach, 1959
(courtesy of Stanley Auerbach)*

Wolfe's direction the AEC began a decade of increasing support for ecological research at commission laboratories, including Oak Ridge, and at universities.

At Northwestern Auerbach had studied the ecology and taxonomy of forest centipede communities. He had also gained experience in research organization, running Park's summer field sessions for several years. After graduation he obtained a position at Roosevelt University. But after four years of teaching with little time for research, he was eager to accept the offer from Oak Ridge. He had also become interested in the AEC's program, even applying for a position at the Argonne National Laboratory, where he was told they had no need for ecologists.[11] Auerbach's appointment thereby demonstrated the importance of a preexisting interest in ecology among health physicists in establishing the discipline at Oak Ridge. The contrast with some other national laboratories, including Argonne, that were much slower to establish ecology programs, is apparent. However, a similarity between the origins of the Oak Ridge ecology program and the Nature Conservancy is also apparent. Although support from the Cabinet's Scientific Advisory Committee was probably less crucial to British ecologists than was sponsorship of ecology by Oak Ridge health physicists, both episodes demonstrate the importance of support from more established disciplines in formulating an ecology program.

That ecology became established at all within the AEC may appear paradoxical. Ecology was often considered "soft," or not fully scientific, particularly by the physicists and other "hard" scientists that dominated the AEC. The AEC's initial support of ecology has been attributed to concern about radioactive contamination of the environment.[12] But in the early 1950s this concern and the perception of the need for ecology were certainly not widespread within the AEC.[13] At Oak Ridge, ecology gained a foothold not by being recognized as an independent discipline relevant to AEC problems but because two health physicists—who were established within the laboratory—considered it essential to their work and defined it as a component of their own specialty.

Auerbach's initial research was entirely consistent with the view of Morgan and Struxness, who considered ecology a component of health physics. In one project he measured radiation effects on laboratory populations of the arthropods found in tree holes. He had already studied similar populations at Northwestern. He continued these studies until 1957, broadening them to include other soil-dwelling arthropods, as well as bacteria.[14] Auerbach also began in 1955 to study the movement of Strontium 89 through earthworms, plants, and soil. These studies of the effects of radiation on soil organisms, and the organisms' movement of radioactive materials, were demonstrably relevant to the waste disposal program. At this time, low-level wastes at Oak Ridge were pumped into unlined earth pits and allowed to seep into the surrounding soil. The wastes were expected to bind with soil particles, immobilizing them.[15]

Throughout the 1950s Auerbach worked closely with health physicists, who provided him with advice on radionuclides, dosimetry and soil geochemistry, as well as data on environmental contamination. Morgan, Struxness, and Auerbach also developed a long-term plan for ecological research at Oak Ridge.[16] Auerbach duplicated in his own research the dual focus of health physics on the biological effects of radiation and the environmental movement of radioactive substances.

Health physics priorities were also evident in Auerbach's shift in 1956 from laboratory to field study. In 1955 White Oak Lake was drained. It had been used since 1943 as a disposal site for the laboratory's low-level radioactive waste, and it now was an area of contaminated land virtually unique in the

United States. To health physicists it made possible far more realistic study of radioactive contamination than did laboratory experiments. Struxness, accordingly, urged Auerbach to shift his focus to the lake bed. So did Wolfe, who told Auerbach that AEC management did not consider his laboratory studies relevant to commission concerns. Within the AEC generally, field studies of radioactivity had become a higher priority because of the difficulty in simulating contamination in the laboratory.[17]

Auerbach, however, was reluctant to move from the laboratory to the field, where it was difficult to set up controlled experiments. Variations in contamination, weather, and other conditions could, he feared, make reproducible results more unlikely. It had been Auerbach's understanding that one factor leading to the termination of Krumholz's work had been the wide variations in his data, which, while perhaps inevitable in field surveys, suggested to Oak Ridge scientists a lack of rigor. However, he had to balance this concern for the perception of ecological research as less rigorous with the need to demonstrate its contribution to understanding radioactive contamination. Accordingly, he began in 1956 to study the movement of radionuclides within the soil, plants, and fauna of the lake bed, and the effects of radiation on the biota.[18] In 1957 he and his colleagues planted corn, legumes, and other crops on the lake bed to measure their uptake of radionuclides and the effect of radiation on their growth. For agricultural experiments the lake bed, which was contaminated by several isotopes, simulated fallout more realistically than did laboratory studies, in which plants were treated with single radioisotopes.[19] Presumably because of this realistic experimental approach to a potential radiation hazard, these studies impressed Alvin Weinberg, director of the Oak Ridge laboratory.

The same year, ecologists studied another site contaminated by radioactive waste. Health Physics Division hydrologists had detected seepage of a radionuclide, Ru 106, from several of the waste pits. To determine whether it had reached the surrounding forest Auerbach and his colleagues measured the radionuclide content of trees near the pits. This study, like those of radionuclide uptake by lake-bed vegetation, demonstrated that ecologists, by evaluating the hazards of radioactive waste disposal into the soil, could contribute to their management.[20]

The contrasting experiences of Ovington and Auerbach as they began their research careers in the early 1950s demonstrate how the status of ecology within their institutions influenced their interaction with neighboring disciplines. Both took advantage of opportunities provided by nonecologists: Ovington studied Forestry Commission plantations, and Auerbach focused on problems provided by health physicists. Ovington, however, was able from the start to pursue work of interest primarily to ecologists, not foresters, while Auerbach's early work addressed health physics priorities. Within the conservancy Ovington was able to address ecological priorities even as he used facilities provided by another agency; the Oak Ridge laboratory, in contrast, was indifferent to ecology, and Auerbach owed his status there to the initiative of health physicists. A further contrast was that while Auerbach was at first reluctant to begin field experiments, because of concern that his results might appear variable and inconclusive to physical scientists, Ovington and other conservancy ecologists had no such reluctance, as their work would be evaluated by other ecologists, who would be more understanding of unavoidably variable results.

But while health physicists provided the initial place for ecology at Oak Ridge, Auerbach began in the late 1950s to broaden his program beyond their immediate requirements. In 1957 he divided his program in two: studies of the environmental movement of fission products relevant to the hazards of radioactive waste, reactors, and weapons fallout; and "long-range" studies of the flora and fauna of radioactive areas, including their community structure and the effect of environmental factors on species abundance.[21] Long-range research would focus less on the movement and effects of radioactive materials than on gaining a basic understanding of ecological phenomena. A further step toward independent ecological research was made in 1958, with studies of forests relatively uncontaminated by radioactive wastes. I discuss this research below.

The following year Oak Ridge ecologists broadened their research further to include the nearby Clinch River. Daniel Nelson, who had just been hired by Auerbach, participated in a cooperative survey of radioactive contamination of the river. He studied the uptake and accumulation of radionuclides by clams and other aquatic organisms.[22]

Auerbach sought greater independence for ecology just as the waste research project, for which he had been hired, was ending. The AEC had little commitment to waste disposal research, and Auerbach's seeking of a role independent of this problem can be seen as an effort to gain greater security for his own program.[23] His aim also was evidence of the ambition of AEC ecologists, including Wolfe and Eugene Odum, who had been associated with the AEC since 1951, to achieve an independent status within the commission. With the AEC now dispensing considerable funds for research, the ecologists sought to ensure that some be allocated to research projects developed by ecologists. The methods and perspectives specific to ecology, they argued, were essential to understanding environmental aspects of radiation. Ecologists could therefore best contribute to AEC goals by pursuing their distinctive research approach.[24] Wolfe, for example, testified to the need for research devoted to ecological goals, stating that "it seems essential that an ecology program be so organized as to utilize techniques and modes of thought peculiar to the point of view [of ecology]. It thus may contribute to the needs of Health Physics rather than attempt to meet those needs by way of a partial approach."[25]

Auerbach made a similar argument, asserting that problems of radioactive contamination of the environment demanded basic ecological research: "Both long-range and emergency problems have a common essential requirement; namely, discovery of the principles governing the maintenance or renewal of biological resources."[26] His ambition, evidently, was to build a program that could contribute to the technological objectives of the AEC, including the development of applications of atomic energy, while providing opportunities for research of interest to ecologists.

Auerbach's initial plans for ecological research emphasized study of species populations and communities. He soon decided, however, in consultation with Odum and after reading Odum's textbook *Fundamentals of Ecology*, to base his program on ecosystem ecology. Through discussion with his health physics colleagues and other scientists he had become convinced that only a quantitative, physicochemical perspective would find acceptance at Oak Ridge. Ecosystem ecology promised such a perspective because it emphasized the movement of energy and materials within self-regulated systems made up of both living and nonliving matter and deemphasized the unique properties of species. Accordingly, he began explaining his program to Oak Ridge and AEC management in terms of the ecosystem approach, his arguments reinforced by Odum, whom he asked to serve on the Division Advisory Committee.[27]

By this time the ecosystem approach had become a significant new field of study within American ecology. Seven years after Tansley had coined the term, a young American ecologist established what many ecologists consider the foundation of ecosystem research. Raymond Lindeman's paper "The Trophic-Dynamic Aspect of Ecology" appeared in 1942, a few months after his death at the age of twenty-seven.[28] In it he discussed concepts long of interest to ecologists, including succession and the significance of feeding, or trophic relationships to the structure of ecological communities, as elaborated by Charles Elton.[29] What was novel was how he integrated these concepts. Energy, he explained, was the common denominator that could relate successional changes to the productivity of trophic levels (green plants, herbivores, and predators, for example) and the efficiency of energy transfer between these levels. In effect, he showed how to relate long-term change in ecosystems to short-term events of food consumption, respiration, and other aspects of the flow and transformation of energy.

Lindeman's perspective was especially significant because it redefined nature for ecologists. They had generally viewed nature as the behavior and interactions of species against the backdrop of an abiotic environment. Succession, for example, was the replacement of one species assemblage by another; study of food relations within a community began with identification of certain species as predators, others as prey. Lindeman suggested instead that an ecosystem be viewed as functional components: trophic levels, not species, were central to ecological analysis. By reducing the complexity of food chains and ecological change to energy flow, he made ecosystems amenable to quantitative physicochemical analysis. Establishing a common basis for plant and animal studies in the movement and transformation of energy was also a step toward a single unified ecology, and a step away from the view, held by Tansley, Elton, and others, of ecology as "scientific natural history" grounded in appreciation of the uniqueness of each species.

By emphasizing the functional roles of ecosystem components, Lindeman also undermined the distinction between living and nonliving components. What, he asked, about a dying pondweed, covered with periphytes? Was it alive or dead? Even after death the plant retained a function as a source of nutrients. With rapid transfer of nutrients between living and nonliving ecosystem components the distinction between them, Lindeman argued, was arbitrary.[30]

Lindeman wrote his paper while a postdoctoral student with G. Evelyn Hutchinson at Yale University. Hutchinson had been considering similar themes in his own research, including the relation between succession and trophic relationships.[31] Like Tansley, Hutchinson would not follow Clements in his view of the community as an organism, and yet he found the analogy suggestive, noting that "if the community is an organism, it should be possible to study its metabolism."[32] Through his reading of Victor Goldschmidt's works on geochemistry and Vladimir Vernadsky's explication of his biosphere concept, his limnological studies of Linsley Pond in Connecticut, and his commitment to the concept of equilibrium, Hutchinson developed a theoretical understanding of the "metabolism" of ecosystems.

Hutchinson outlined his views in a 1946 paper titled "Circular Causal Systems in Ecology." In it he described the movement and accumulation of carbon in the biosphere and phosphorus in lakes. Organisms, he noted, influenced the movement of these elements, and in turn their productivity was in part determined by the availability of these substances. He also developed mathematical equations depicting the growth and interactions of populations.

Underlying both phenomena—the movement of elements and the behavior of populations—were circular causal paths, or feedback loops, damping oscillations and maintaining equilibrium, thereby ensuring the persistence of the system.[33] Hutchinson had derived the feedback concept from recent developments in the study and management of complex systems. Operations researchers, stimulated by the demands of World War II, had demonstrated how complex technology, such as missile guidance systems, could employ feedback loops to ensure optimum performance. After the war the study of self-regulation through feedback, or cybernetics, was transferred to peacetime research, particularly through a series of Macy Foundation conferences held between 1946 and 1953, in which Hutchinson participated.[34]

Hutchinson's students also had considerable impact on ecosystem ecology. One such student was Howard T. Odum. In 1950 Odum completed his dissertation, a study of the biogeochemistry of strontium. Its global distribution, he concluded, had long remained constant; this exemplified the self-regulation and stability of the "strontium ecosystem." In subsequent research Odum measured energy flow between trophic levels in Silver Springs, a series of mineral springs in Florida.[35] He drew energy flow diagrams that he converted into electrical circuit diagrams, using a symbolic language to portray generalized patterns of energy flow. Describing such patterns and reducing ecosystem complexities to flows of energy, Odum believed, would permit discovery of general ecosystem principles. In 1955 he proposed the "maximum power principle," which stated that those ecosystems or other forms persist "which can command the greatest useful energy per unit time (power output)." The practical implication, Odum believed, was that optimum social and ecological organization implied maximum use of energy.[36]

Howard Odum's older brother, Eugene, was also an ecosystem ecologist, but he had a somewhat different perspective on energy flow and biogeochemical cycling. While Howard drew analogies between ecosystems and physical systems, Eugene related the order and stability of the ecosystem to physiological mechanisms of homeostasis. Eugene was also probably more influential among ecologists. At his Institute of Ecology at the University of Georgia, and in three editions of *Fundamentals of Ecology*, Eugene Odum alerted ecologists, including Auerbach, to the potential of ecosystem ecology. The result was a small but growing number of ecosystem studies during the 1950s. These included John Teal's study of energy flow in a salt marsh, Ramon Margalef's application of information theory to ecosystems, and Bernard Patten's work in cybernetic theory.[37] As Ovington's interest demonstrated, the influence of the ecosystem approach was not limited to the United States.

By 1960, then, a growing number of ecologists interpreted nature in terms of ecosystems—as large as the biosphere, as small as a pond—within which energy and nutrients are exchanged, consumed, and transformed, and that possess feedback loops ensuring that the system would remain, within limits, at equilibrium. This common interest of some American and British ecologists was evidence of a strengthening sense of membership in an international discipline of ecosystem ecology; it was defined not only by this shared interest but also by the institutions providing the scientific autonomy and encouragement necessary for such research.

One implication of the ecosystem approach was a shift in ecological metaphor. Clements and other ecologists had viewed ecological communities as organisms, suggesting that they could be understood only in terms of the unique complexity of living entities.[38] Tansley, while rejecting this view and interpreting communities (somewhat vaguely) as "systems," nevertheless retained a focus on species acting within a physical environment. Hutchinson and his students eliminated the barrier between the study of living and nonliving nature. The functioning of ecological systems was not, in principle, distinguishable from that of physical systems; rather, living and nonliving systems obeyed common mechanical principles. Both types of systems tended to remain at a point of stable equilibrium and to return to this point when perturbed, and regulatory processes maintaining this equilibrium could be understood and possibly manipulated to ensure the system's optimum behavior. Both types of systems, in other words, are cybernetic. Nature became less an organism than a machine.

This approach to ecology readily accommodated the expectations of physical scientists at Oak Ridge. It also helped Auerbach assert a more distinct identity within the laboratory. Research activities at Oak Ridge were organized by divisions. Among nonecologists at Oak Ridge, ecology was considered most closely allied to the Health Physics Division or to the Biology Division. While it had originated within health physics, Oak Ridge management, at least initially, thought that ecology might better belong within the Biology Division. But Auerbach argued that ecology was distinct from both. The ecosystem, invoked by some ecologists as their distinctive unit of study but unfamiliar to both health physicists and biologists, supported this distinction. It nevertheless also suggested that ecology complemented biological and health physics research. Oak Ridge biologists studied the effects of radiation on cells and selected organisms but without considering the pathways by which radiation was delivered to the organisms.[39] Health physicists were concerned with the exposure of humans to radiation but did not consider the role of the ecosystem in mediating this exposure. Therefore, ecosystem ecology complemented the work of both—by its focus on processes affecting the interaction between radioactive materials and humans or other species—even as it asserted its distinctiveness. While ecology's initial status as a component of health physics had given it a foothold within the laboratory, after five years Auerbach considered its position secure enough to permit a more independent identity. His strategy toward health physics and biology, in other words, evolved into one analogous to that adopted by the conservancy toward agriculture and forestry: the demonstration that ecology could complement the work of closely related disciplines. In both cases the ecosystem concept was an important element of this strategy, proclaiming the distinctiveness of ecological study.

Adherence to a physical science model, distinctive objectives, complementarity with neighboring disciplines; this was the balance required in the fostering of a new, independent discipline among others already entrenched within the laboratory. For Auerbach, the ecosystem approach permitted ecology to achieve this balance, and in doing so, flourish at Oak Ridge. By 1960 his ecology program had become one of the largest in the United States, with twenty-two permanent and temporary employees and visitors (fig. 5). They pursued a variety of projects, on sites of varying geology, level of radioactive contamination, and flora and fauna, mostly within the 15,000-hectare AEC reservation surrounding the laboratory. This research also reinforced Auerbach's argument that ecology merited a place at Oak Ridge. Just as nuclear physicists, for example, justified their presence at Oak Ridge on the basis of its reactors and other equipment, a similar case could be made for ecological re-

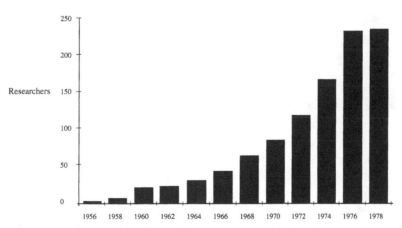

Fig. 5. *Researchers in Oak Ridge Ecology/Environmental Science Program,*
1956 to 1978
Source: *Oak Ridge National Laboratory annual reports*
Note: *Figures include permanent and temporary staff and visitors*

search because of its need for large areas of undisturbed land: "The Oak Ridge Reservation constitutes a large-scale field research facility, analogous to the reactors, cyclotrons, and chemical and biological laboratories of other groups."[40] The combination of facilities and environmental problems of nuclear technology, were, Auerbach argued, a unique opportunity for ecology.[41] As the program grew it gained a more independent status within the laboratory. In 1959 it became the Radiation Ecology Section of the Health Physics Division. Two years later it moved into its own building, nearly tripling the space available for ecology.[42]

RADIOECOLOGY AND ECOSYSTEM ECOLOGY

While Auerbach developed an ecological perspective compatible with the constraints and opportunities of Oak Ridge, translating that perspective into research was the work of the ecologists he attracted to the laboratory. While they shared certain approaches, such as the study of the movement and accumulation of radionuclides within ecosystems, their research encompassed a variety of organisms—insects, mammals, and plants—and strategies that reflected their diverse experience and interests.

D. A. Crossley, an entomologist trained at the University of Kansas, became in 1956 Auerbach's first recruit. At Oak Ridge he explored the interaction between arthropods and plants, including changes in lake-bed arthropod populations during succession, the movement of radionuclides in plant-insect food chains, and the role of invertebrates in forest litter decomposition. He also developed a technique, using radionuclides, for measuring the consumption of plant material by insects. It could also, he argued, be used in assessing the risk to humans of the transfer of radioactive materials by insects.[43] The insects, however, took up only a tiny fraction of the lake-bed radioactivity, and they were outside any food chain leading to humans. Clearly, his interest in using radionuclides was to study food chains and other phenomena that had long interested ecologists, and not to evaluate a potential radiation hazard.

In 1967 Crossley left Oak Ridge for Odum's ecosystem program at the University of Georgia. Some of his research, however, was continued by David Reichle, who had arrived in 1964. Reichle studied aspects of forest nutrient cycles, including arthropod food chains, the role of bacteria and insects in litter decomposition, and the rates of nutrient cycling in different soils. He and Crossley also showed how the use of radionuclides simplified measurement of food consumption, thereby aiding study of the energetics of food webs and ecosystems. Reichle also explored various implications of the movement of radioactive contaminants in the environment.[44]

In December 1957 Paul Dunaway began study of the population dynamics of small mammals of the lake bed and surrounding area. Stephen Kaye joined him in 1960, in a long-term study of the accumulation and elimination of radionuclides by these mammals, and the biological effects of environmental radiation. Seeking to explain why vertebrate species vary in their sensitivity to radiation, they focused on the effects of radiation on the tissues responsible for the formation of blood and blood cells.[45]

In 1958 Jerry Olson came from the Connecticut Agricultural Experiment Station to Oak Ridge to study forest and soil ecology. As a student at the University of Chicago he had studied plant communities of the Indiana Dunes, receiving his doctorate in 1951. This research on dunes was noteworthy for his mathematical perspective on the circulation and chemical transformation of soil compounds and nutrients during succession. It reflected his view of plants and soil as a single "dynamic physical-biological system," a perspective partly inspired by Henry Chandler Cowles.[46]

At Oak Ridge Olson applied radiological and computing methods to his studies of soil and plant ecology. In April 1959 he and several colleagues began to study forest nutrient cycling. In a series of experiments they injected radionuclides into one or a few trees and then tracked the radionuclides' movement. By assuming that radionuclides and nutrients circulate in similar ways, Olson inferred the movement of nutrients into the soil as litter decomposes, their subsequent uptake by trees, and their eventual return to the soil. Through studies in varying conditions he compared the effects of tree species, climate, soil fauna and other factors on litter breakdown and element circulation.[47] In 1962 Olson and his colleagues began their largest study: the injection of cesium 137 into thirty-five tulip poplar trees (*Liriodendron tulipifera*). This "cesium forest" remained for several years the focus of forest research at Oak Ridge. As in earlier studies the ecologists inferred from the movement of the cesium information about nutrient cycling and primary productivity. Related experiments on arthropods within the forest focused on their food consumption and secondary productivity.

Olson's use of cesium was significant. Although it was easier to use and interpret than other elements, it also distinguished his study from those of many other AEC researchers who, because of widespread concern about weapons fallout, were focusing on the environmental behavior of strontium 90.[48] The suitability of cesium for research clearly outweighed the relevance of strontium to concerns about environmental contamination. Its use reflected Olson's primary goal of understanding the functional processes of ecosystems; this understanding could then, he argued, be applied to understanding radionuclide movement.[49]

Joel Hagen has explained how, after World War II, ecologists found that the use of radionuclides as tracers greatly aided in quantifying the movement of materials and energy through ecosystems.[50] As research tools, radiotracers were considered far superior to the uncertain quantities and variety of radionuclides contaminating the lake bed and waste pits. Besides easing replication of experiments, radiotracers made possible radioecological research anywhere in the AEC reservation, not just near the waste sites, where the complexity of the vegetation and their contamination made results more difficult to interpret.[51] Radiotracers, therefore, not only had the methodological advantages that Hagen identifies, they also effectively broadened the geographic scope of radioecology at Oak Ridge by making possible greater replication of experiments. But in releasing radioecology from the waste disposal areas they also lessened its relevance to the concerns of health physicists.

While applying radiotracers to forests Olson also experimented with computer models.[52] Beginning in September 1958 he developed a series of models in which components of the forest ecosystem, such as soil, litter, and trees, were represented as compartments linked by equations simulating the movement of elements between them. These models became progressively more complex; by 1963 he had developed a nine-compartment model able to simulate the movement of carbon, nutrient elements, or radionuclides. Olson also linked his modeling with the cesium forest experiment. By comparing model predictions with the movement of cesium in the forest itself he could identify shortcomings in the model.[53]

Olson considered models, like radionuclides, to be most useful as tools for ecological research. By simulating the essential processes of nutrient cycling and productivity, radionuclides served as workable surrogates of an actual ecosystem, making it possible to consider the characteristics of the entire system, not just its parts: "Mathematical analysis of schematic models or 'circuit diagrams' of ecosystems is more than an expedient device for expressing the movement of particular radioactive contaminants from one compartment of the system to another. The parameters describing the rates of cycling of major and minor elements through the system express the fundamental characteristics and organization of the system as a whole."[54] Models could, for example, aid in the study of aspects of forest productivity and succession too slow to observe directly. A model simulating the accumulation and decomposition of litter on the forest floor could predict when these gains and losses would be balanced and a climax steady state attained. Olson reformulated other aspects of succession to make them amenable to simulation modeling.[55] His research, therefore, reflected a continuing effort to apply the tools available at Oak

Ridge—models and radionuclides—to ecological phenomena that had been of interest to him since his studies at Chicago.

By 1964, Crossley, Reichle, Dunaway, Kaye, and Olson were engaged in field, laboratory, and modeling studies of the movement and effects of radioactive substances in the terrestrial environment. They and Auerbach presented their results as empirical knowledge of the behavior of radionuclides that could be applied to prediction and management of radioactive contamination of the environment. But while using radioactive substances they also pursued such areas of interest as insect ecology, forest nutrient cycling, and physiological aspects of radiation damage, that extended beyond understanding the environmental aspects of radiation. Their research contributed to the growth of radioecology, a field that, as described by Odum, Auerbach, and other prominent AEC ecologists, sought to bridge the gap between basic ecological research and the nuclear sciences.[56] Auerbach and other Oak Ridge ecologists presented radioecology as potentially useful in dealing with radioactive contamination of the environment. However, its primary focus was the application of tools made available by the AEC to problems of long-standing interest to ecologists, such as the characteristics of food chains and the processes of succession.

By the early 1960s some AEC ecologists had also decided to strengthen ties between radioecology and ecological theory. Odum, for example, suggested that ecologists be less preoccupied with development of radionuclide techniques and empirical study of radionuclide movement, and instead apply these techniques to building ecological theory.[57] Auerbach bolstered his section's capability in theoretical ecology by hiring Bernard Patten in 1963 and George Van Dyne the following year. Although Patten stayed only five years and Van Dyne just three, both added to the laboratory's reputation for ecosystem theory and modeling.

Patten had been a marine biologist at the College of William and Mary and at the Virginia Institute of Marine Science. While his experience was not immediately relevant to the forests and rivers near Oak Ridge, Auerbach saw in Patten's studies of the cybernetic behavior of ecological systems the mathematical and theoretical sophistication that he wanted to encourage.[58] At Oak Ridge Patten studied models of competition between populations stressed by radiation, the development of equilibrium states in ecosystems receiving radioactive materials, and the movement of radionuclides within a laboratory microecosystem.[59]

Van Dyne had studied animal husbandry at Colorado State University and at South Dakota State University, and nutrition at the University of California at Davis. His agricultural background, like Patten's marine experience, was not relevant to the problems preoccupying Oak Ridge ecologists. However, Auerbach knew of Van Dyne's interest in mathematical approaches. And Auerbach was planning a large experiment in which several hectares of pasture would be contaminated with simulated radioactive fallout. Van Dyne, with his knowledge of grassland studies, could contribute to research. An interest in Olson's ecosystem models drew Van Dyne to Oak Ridge.[60] Once there, he immersed himself in the opportunities provided by the national laboratory, writing close to two papers per month, a pace that "shook the rest of the staff completely."[61] In these papers he experimented with various computer modeling and systems analysis techniques, such as operations research, building an understanding of how comprehensive ecosystem studies could be conducted. The ideas he would develop at Oak Ridge would find expression in the 1970s, when he directed the International Biological Program's (IBP) Grasslands Biome project.[62]

Patten's and Van Dyne's work owed little to practical problems of radioactive contamination or to the field studies being conducted at Oak Ridge. In their models they often chose experimental and interpretative tractability over precise empirical prediction. Although Patten would mention radiation or radionuclides in his research, they usually were used as examples of more general phenomena. In his studies of the effects of radiation on competition between two species, for example, he introduced these effects into Lotka-Volterra equations as a stress resulting in reductions in each population, following the standard format employed for other stresses such as predation or harvesting.[63] Patten was more interested in the application of mathematical theory, such as network theory, to ecology. Ecologists, he believed, should base their research on a foundation of mathematical theory. This was illustrated by the course in systems ecology that he, Olson, and Van Dyne taught for several years at the University of Tennessee. Its emphasis on mathematical principles was intended, Patten wrote, to develop in students an "appreciation for the natural affinities between ecology and mathematics."[64] Neither Patten nor Van Dyne relied on empirical data from other Oak Ridge research. One of Patten's largest studies was based on data collected while he was a marine biologist in Virginia, and Van Dyne relied largely on data from western grasslands, possibly because the field experiment that had helped attract him to Oak Ridge was delayed by budget restrictions.[65]

Patten, Van Dyne, and Crossley had left Oak Ridge by 1967, and Auerbach and his colleagues, eager to maintain a reputation for innovative ecosystem research, had started new projects. That year Daniel Nelson and James Curlin chose the Walker Branch, a stream near the laboratory, as the site for a watershed ecosystem study. They did so just as American ecologists were expressing greater interest in such studies. F. Herbert Bormann and Gene Likens had published the first results of their Hubbard Brook study, and the previous fall ecologists had decided to base much of the U.S. International Biological Program on a set of watershed studies. The Walker Branch study could also make use of the scientists' individual expertise: Nelson had studied stream ecology as a student, and Curlin was a professional forester.[66]

Nelson and Curlin intended to focus on the stream itself, not the entire watershed. While Bormann and Likens were showing that a watershed could be studied by comparing the streamflow leaving it with the precipitation entering it, this was not possible at Walker Branch because the bedrock was not watertight. Nelson and Curlin nevertheless argued that stream research should encompass its watershed, because the stream continually imports both energy and elements from the surrounding land, which influence stream chemistry and biological productivity. They also had a practical justification for this study, which they defined in terms of the entire federal water research effort. The study, they explained, could provide a unique perspective on pollution and other aquatic environmental problems, because while several hundred experimental watersheds had been studied over the years by the Forest Service and other agencies, few studies shared their focus on water quality.[67]

This justification, while remote from AEC missions, was nevertheless consistent with Weinberg's priorities. The place of national laboratories in federal science was changing. As reactors moved from the experimental laboratory into commercial production, debate grew over whether massive nuclear research facilities were still needed. The makers of federal energy policy were beginning to consider a broader range of options, of which nuclear energy would be only one. Thus, nuclear research was likely to receive, as Weinberg perceived, a diminished share of energy funding. In response, he argued that the laboratory must develop new missions in order to remain relevant to changing societal priorities, and he supported new, nonnuclear projects for his laboratory, such as the desalinization of seawater. He also sought non-AEC support for Oak Ridge research, lessening the laboratory's dependence on a shrinking commission budget. By 1969, 14 percent of the research at Oak Ridge was sponsored by outside agencies.[68]

Weinberg also envisaged a larger role for Oak Ridge in environmental research, and he encouraged Auerbach to help make this a reality.[69] In approving his colleagues' plan for Walker Branch, Auerbach was following the direction Weinberg had established for the entire laboratory. A 1967 amendment to the Atomic Energy Act, which gave the AEC authority to support environmental research, facilitated such initiatives.[70]

But while Walker Branch was a step into nonnuclear research, other Oak

Ridge ecologists were becoming involved in studies relevant to radiation hazards. Kaye turned from the study of small mammals to the environmental movement of radionuclides hazardous to humans, contributing to an assessment of the risks of using nuclear explosives to dig a new Panama Canal. This canal was to be a centerpiece of the AEC's Plowshares Program, which aimed to develop peaceful applications for nuclear weapons. Kaye and Sydney Ball of the Oak Ridge Instrumentation and Controls Division combined an environmental compartment model with predictions of the radiological dose to humans to predict the radiation risks to communities near the proposed canal. This research differed from other ecologists' studies of radionuclide movement by focusing not on one radionuclide (as in studies of the cesium forest, for example) but on twenty-five that were likely to be produced by nuclear explosions. In addition, the model was designed not to simulate cycling of radioactive materials within each component of the environment but to predict their accumulation in a single compartment: humans.[71]

In 1968 several Oak Ridge ecologists began another study of a potential radiation hazard. They spread simulated nuclear fallout, tagged with cesium 137, over a meadow and then followed its movement. The AEC's Office of Civil Defense supported this project, because its primary objective was to evaluate radioactive contamination that could result from nuclear war. But the ecologists also used it as an opportunity to study insect community dynamics, vegetation succession, the role of soil organisms in decomposition, and other ecological topics.[72]

By 1968 the Oak Ridge Radiation Ecology Section was one of the largest ecological research groups in the United States. Ecologists pursued a variety of projects, some directed toward potential radiation hazards, such as those posed by the Plowshares program. Others, such as the cesium forest study, epitomized the combination of radiological techniques and ecological objectives characteristic of radioecology. Still other projects, such as the Walker Branch study, were largely independent of radioecology or AEC missions.

The development of ecological research at Oak Ridge was the result of ecologists' accommodating both the constraints and opportunities of the laboratory context. Auerbach, like other politically astute research directors, considered his chief objective to be the continued expansion of his program. This required, at the minimum, demonstration that ecology belonged at Oak Ridge. It had to be "respectable" in the eyes of the physical scientists (and a few biologists) who directed Oak Ridge and who often, Auerbach recalled, viewed ecologists as little more than a "bunch of butterfly chasers."[73] As I have noted, ecosystem ecology provided the necessary respectability.

Beyond this, Auerbach and his colleagues employed several strategies to demonstrate that ecology belonged at Oak Ridge and within the AEC. One major AEC mission was the dissemination of radiological methods into the scientific community. This was part of its continuing effort to convince American society of the benefits of atomic energy through, for example, its Atoms for Peace program. Oak Ridge ecologists contributed to this mission, in part through summer sessions in radioecology. Beginning in 1962 they instructed visiting scientists in the use of radiological methods in ecology. But a wider dissemination of these methods was achieved through their own research. By publishing results obtained using radiotracers, Auerbach and his colleagues could show that such techniques had practical research applications. As early as 1958 Auerbach had noted the use of radionuclides in tracing the flow of energy through the soil ecosystem.[74] Later he emphasized the relevance of radiological methods to ecology, describing radionuclide cycling as part of the general problem of element circulation in ecosystems: "Radioactive tracers offer almost unexplored opportunities for investigating ecological processes in the landscape. Nutrient and trace-element cycling in ecosystems, delineation of food-chains, or any other ecological process which involves the transfer of material between two components of an ecosystem, are subject areas where radioactive tracers may be useful. Perhaps even more important, tracers provide the ecologist with a tool for tagging the entire biota of a community. By so doing the ecologist is in a position to estimate the interactions between the components of a community."[75] Radiotracers not only aided research by providing more control over experiments, they also provided a political benefit by helping to demonstrate that ecology could contribute to an important AEC mission. This benefit was accentuated by the fact that it was at Oak Ridge that radioisotopes were first developed for general use in research.

The ecology program may also have gained acceptance at Oak Ridge by being organized around a series of large projects. Weinberg believed that national laboratories should be dedicated to big science—projects that required many scientists, whether because of the scale of the research equipment or the complexity of the problem.[76] The largest research divisions at Oak Ridge (high-energy physics, nuclear physics, and biology) epitomized this view. By 1962 Auerbach realized that ecology could also become big science, as the cesium forest experiment indicated. The Walker Branch and nuclear fallout studies extended this strategy.

Simulation models provided a somewhat similar benefit to Auerbach and his colleagues. The ecologists' use of models demonstrated that the methods of ecology were similar to those of other disciplines at the laboratory. Referring to the use of models by health physicists to predict the movement and accumulation of radionuclides in humans, Olson explained that "mathematical treatment of compartment models is used increasingly for the physiological or biophysical analysis of movement of materials between cells or between organs of a single individual"; its application to ecosystems was simply a "logical extension of this methodology."[77] Computer models were also applied to the study of other complex systems, such as nuclear reactors. By using models Olson aimed to demonstrate that ecosystems were essentially equivalent to the engineered systems that justified the existence of Oak Ridge itself: "The basic operation of the analog computer is to keep a running balance of . . . gains and losses for all the major parts or 'compartments' of a system, be it a reactor or an ecosystem."[78]

Olson's parallel between reactors and ecosystems reflected the assumption inherent in early Oak Ridge simulation models, and, as I have noted, in ecosystem theory generally at this time, that an ecosystem is self-regulating, or cybernetic. Chunglin Kwa has suggested that this perspective could gain acceptance among scientists and government officials inclined to view nature and society as systems amenable to expert analysis and management.[79] And as Olson's analogy between a reactor and an ecosystem suggests, such a view may have been especially acceptable within the AEC, an agency whose primary mission was the design and control of complex technological systems. However, caution is needed in evaluating the role of simulation models in ensuring the acceptability of ecology at Oak Ridge, as Weinberg was himself skeptical about such models.[80]

Before 1970, therefore, Auerbach and his colleagues pursued a variety of strategies to win acceptance at Oak Ridge: contributing to AEC missions; complementing but not overlapping the research of neighboring disciplines; and developing projects epitomizing big science. Most important, ecology had to be the kind of science—quantitative, using computers and radiological techniques available at Oak Ridge, and built on physical and chemical principles—that would meet the expectations of physical scientists. A research program that viewed the environment in terms of ecosystems defined by flows of matter and energy fulfilled these requirements.

Similarities between Oak Ridge research and ecology at Merlewood illustrate how studies of ecosystem processes, including nutrient cycling, were not simply the product of institutional opportunities at the national laboratory. While employing quite different approaches, the problems addressed by Olson and his group, and by Ovington and his colleagues, were often similar: for example, both sought to quantify the movement of elements between trees and soil. At the same time, these different approaches illustrate how ecosystem research strategies could diverge, given differences in ecologists' backgrounds, interests, and resources. British ecologists, lacking contact with atomic energy research, did not have radionuclides and simulation models available to them; nor did they have the institutional incentives to use these methods, as Oak Ridge ecologists did. However, when presented with the opportunity in America, Ovington did experiment with the use of radionuclides.[81]

As I have indicated, Auerbach had by the late 1950s identified the ecosystem approach as the one best able to gain acceptance at Oak Ridge. But having chosen to build such a program, he did not direct scientists into ecosystem ecology but hired those already interested in this approach. He then gave them considerable latitude, confident that in following their own interests they would contribute to the larger objective. Such an approach was possible because ecologists were able, within the constraints I have noted, to determine their own research. After the end of the waste disposal program for which Auerbach was originally hired, they were rarely called on to deal with incidents of radioactive contamination or other problems encountered by the Health Physics Division. They did present much of their research as potentially relevant to managing radioactive contamination of the environment. But because their efforts were not directed to specific problems they could match their research to their own interests and priorities as ecologists. Often, as in Olson's studies of forest nutrient cycling and productivity, Patten's explorations of mathematical ecosystems, and Nelson and Curlin's watershed study, the research incorporated interests the scientists had had before coming to Oak Ridge. Overall, the diversity of their techniques and topics, and their frequent lack of reference to actual problems of radioactive contamination of the environment, reflected the latitude they enjoyed to pursue their own ideas.

The autonomy of the ecology program was evident in its growing geographic range. From the contaminated lake bed unique to Oak Ridge to the deciduous forest common to the southern states to the watershed and pasture studies relevant to environments throughout the country, the largest projects exhibited an increasing independence from the local environment at Oak Ridge, and the requirements of the Health Physics Division. Instead, they sought to develop theory and methods of interest to ecologists anywhere.

This correlation between seeking general ecological understanding and independence from the local environment recalls the dichotomy established at Merlewood between research by Ovington and his colleagues on ecosystem nutrient cycling and productivity, and studies of specific problems encountered in management of a nearby nature reserve. At both centers ecologists could not readily contribute to specific practical problems while contributing to basic research. At Oak Ridge, the absence of a requirement to assist in specific local environmental problems left ecologists free to contribute to issues of general theoretical interest. At Merlewood, the divergent needs of local problems and of general ecological research led to largely separate programs of study.

The organization of AEC research helped foster ecologists' autonomy. Auerbach reported to the Environmental Sciences Branch of the Division of Biology and Medicine and was in turn responsible for all ecological research at Oak Ridge. This gave him leeway to select a staff that could fulfill his plan for ecosystem ecology. He could also pursue support from outside the Health Physics Division, such as the Office of Civil Defense, to supplement his budget. Overall, Auerbach's autonomy was consistent with the independence that, until the late 1960s, AEC laboratories traditionally accorded their researchers.[82] In 1970 the Radiation Ecology Section of the Health Physics Division became a separate Ecological Sciences Division. This merely made official the independence that Oak Ridge ecology, under Auerbach's direction, had already achieved.

Another factor that gave Oak Ridge ecologists more autonomy was the AEC's lack of significant concern over radioactive contamination of the environment. Until the late 1960s the Joint Committee on Atomic Energy, which oversaw the AEC's activities, virtually never mentioned ecological research or environmental concerns in its deliberations. Nor did the Truman and Eisenhower administrations exhibit serious concern.[83] Both administrations, at any rate, were willing to leave detailed oversight of atomic energy to the AEC and the Joint Committee, in keeping with the iron triangle corporatist structure of nuclear policymaking. The AEC also occasionally consulted informally with other U.S. agencies, including the Fish and Wildlife Service and the Public Health Service, regarding the environmental impacts of nuclear power plants.[84]

During the 1950s and 1960s there was extensive internal debate within the AEC concerning reactor safety and radiation protection. Such matters were within the domain of the AEC's Advisory Committee on Reactor Safeguards and the Hazards Evaluation Branch of the Division of Licensing and Regulation.[85] Ecological expertise was not usually considered relevant. Ecologists had apparently little contact with these debates, which focused largely on issues relating to engineering and physical science. Similarly, the National Advisory Committee on Radiation, established in 1958 by the surgeon general, included among its members physicians, public-health officials, geneticists and other scientists, but no ecologists.[86]

Nor did ecologists play a significant role in studies of waste disposal. While controversy over waste disposal did lead to research, the focus of attention was on high-level wastes. Research focused on their containment, and required expertise in engineering, chemistry, and geology, but not ecology. Management of low-level wastes, in contrast, could require ecological expertise because they were released into the environment. However, as George Mazuzan and J. Samuel Walker document, the AEC saw its strategy of "dilute and disperse" as safe and environmentally harmless. Overall, low-level wastes were seen less as a technical problem requiring research than an economic and public relations issue.[87]

Thus, with ecology and environmental problems of low priority and visibility, Oak Ridge ecologists were rarely called on for research applied to specific problems. As a result, they had considerable autonomy in defining their research strategies and in explaining the strategies' potential relevance to environmental problems. Nature Conservancy ecologists, as I indicated in the previous chapter, also had much autonomy in defining their relevance, but for a very different reason. The conservancy was an independent institution led by ecologists who could define its mandate. Not surprisingly, the issues it addressed were those to which ecological expertise would be highly relevant. At Oak Ridge, ecologists were a subordinate component of a larger institution that generally defined its problems in ways that minimized the relevance of ecology.

More generally, the autonomy of ecological research at Oak Ridge reflected not only lack of concern about environmental problems but the predominant view of the role of science in American society. In the two decades after World War II the overt assumption guiding American science policy was that scientists should be permitted to set their own objectives and criteria for evaluating research. Ever since the publication of Vannevar Bush's report *Science, the Endless Frontier* (1945), spokesmen for the scientific community had stressed that American scientific leadership depended on undirected basic research.[88] (As we saw in chapter 2, a similar view was current in Britain at the time of the conservancy's formation.) The AEC, led by many of the same physicists who articulated this national science policy, provided substantial support for basic research. It is tempting, therefore, to interpret the AEC's support of ecological research as a manifestation of a commitment to undirected, basic research.

The contexts of science policy at the conservancy and at Oak Ridge differed in an important respect, however. As Paul Forman has demonstrated, postwar American physical science, while ostensibly guided by the ideal of undirected basic research, was in fact shaped by the demands of national security through the massive support provided to it by the military and the AEC, which in return expected to achieve certain technological goals.[89] To some extent the development of ecological research at Oak Ridge also reflected the demands of an overarching technological and political objective. Certainly, ecologists were rarely called on to help solve specific problems, as such problems were a low priority within the AEC. Ecologists did not gain support within the AEC on the basis of their field's relevance to practical problems of environmental contamination. But ecology nevertheless gained stature at Oak Ridge by contributing to a central objective of the AEC: the development of atomic technology. In the atmosphere of optimism prevalent until the late 1960s, in which the AEC sought to help provide a technological basis for American society, fueled by unlimited nuclear energy, ecologists, as Auerbach realized, could benefit most not by elucidating the potential hazards of this vision, but by helping bring it about, demonstrating, in their own discipline, the advantages of the technological and organizational innovations it promised. While ecologists could choose specific research objectives, the larger AEC objective helped determine what choices were available, thereby shaping the development of ecology at Oak Ridge and helping to ensure a prominent place for ecosystem ecology within the American scientific community.

Between 1968 and 1974 several environmental and energy initiatives swept through national affairs. Three of special significance for ecologists were the International Biological Program, for which Congress approved funding of the U.S. component in 1968; the passing in 1970 of the National Environmental Policy Act; and the replacement in 1974 of the AEC by agencies taking a more comprehensive approach to energy. These initiatives were the result of changes in national priorities and in the political context of environmental affairs. Together they would profoundly affect the Oak Ridge ecology program.

5

Oak Ridge Ecosystem Research and Impact Assessment

Nuclear power has played a prominent role in the American environmental movement. In the late 1950s and early 1960s, concern about radiation focused on fallout from weapons tests. Although the 1963 Partial Test Ban Treaty quieted this concern, it reappeared by the end of the decade. With a major expansion of nuclear energy under way, the concern now was over radioactive emissions from power stations and the risks they posed to human health.[1] John Gofman, a physician and nuclear chemist, and Arthur Tamplin, a biophysicist, both at the AEC's Lawrence Livermore Laboratory in California, were at the center of this controversy. They argued that the biological effects of radiation were worse than the AEC suspected: that exposing the American population even to permissible doses of radiation would result in thousands of additional cases of cancer every year. The resulting controversy did not directly involve ecologists, because it turned on interpretation of studies of single organisms removed from their environment. Thus, while it has been said that the Age of Ecology began with the first test of the atomic bomb, the relevance of ecological expertise to the nuclear age was in fact not always evident.[2]

But ecologists could become involved indirectly because of their role in tracing the pathways by which radiation could reach humans. This was the case in the late 1960s, when the AEC's authority over power plant radiation emission standards was contested. Critics argued that by treating each reactor in isolation the AEC was ignoring the combined effects of multiple sources, as well as the potential harm to organisms other than humans. Court battles occurred over various environmental aspects of nuclear power plants. Several states attempted, by imposing tighter regulations, to overrule AEC reactor radiation standards. While unsuccessful, such challenges did prod the AEC to impose new, stricter limits on power plant radiation emissions while contributing to growing opposition to nuclear power.[3]

Aside from their fears about radiation, critics of nuclear power were especially concerned about how power-plant cooling systems would affect aquatic organisms. All thermoelectric technologies require water for cooling the steam used to turn turbines. Nuclear power plants, however, have a uniquely large requirement because of their size, the absence of smokestacks that carry away heat, and, most important, the inefficiency of their conversion of heat energy to electricity. As a result, they produce enormous quantities of heated water. More than any other aspect of nuclear power had before, this issue called on ecological expertise. By 1970 ecologists had found that thermal pollution could affect aquatic organisms in several ways: by decreasing the amount of dissolved oxygen in the water but increasing organisms' demand for oxygen; by disrupting spawning and reproductive cycles; and by changing the composition of aquatic communities.[4] Ecologists and other scientists were also active in local controversies. For example, opposition to a proposed station on Cayuga Lake in New York came originally from scientists at the Water Resources Center at Cornell University. As a result of such activities, thermal pollution became controversial in the late 1960s and early 1970s and was a factor in cancellation of construction or modification of several plants.[5]

The AEC responded slowly to concerns about the environmental effects of nuclear power. For several years it resisted taking responsibility for thermal pollution, insisting that its "regulatory interest is . . . restricted to public health and safety questions relating to specific (radiation) characteristics of nuclear materials and atomic energy."[6] Instead, the AEC and its supporters elsewhere within the government insisted that nuclear power was not a hazard but a solution to environmental problems. As the members of the Joint Committee on Atomic Energy (JCAE) argued, "Those members of the general public with genuine questions and concerns will come to realize that, in terms of their relative impact on the environment, nuclear plants in many respects are the least offensive of the various thermal generating units."[7] They argued that, in fact, nuclear energy could solve major environmental problems by displacing dirtier energy technologies and conserving fossil fuels. A "complete switch to nuclear energy would be an important step in restoring our air to the quality it had before the Industrial Revolution," argued Alvin Weinberg, director of Oak Ridge National Laboratory.[8] This argument convinced some; orders for commercial nuclear reactors surged in the mid-1960s, in part reflecting concern over the environmental effects of coal-fired plants and strip-mining.[9]

Proponents of nuclear power also resisted public discussion of the issue. As Congressman Chet Holifield of the JCAE once remarked, "I am not really sure that the presence of a lot of environmental dilettantes and their hovering legal eagles has ever contributed much to nuclear energy or ever will." The Joint Committee shifted its role from overseer of the AEC's activities to promoter of the commission and atomic energy. It sought to keep nuclear energy development insulated from political controversy through the corporatist structure of atomic energy decision-making—the iron triangle of the JCAE, the AEC, and the nuclear industry.[10]

Even as AEC officials in Washington tried to insulate nuclear power from environmental concerns the commission began expanding its stock of environmental expertise. Brian Balogh has argued that the AEC did so in case it failed in its primary strategy: avoid responsibility for most environmental problems.[11] But, as in other large science-based agencies, AEC priorities did not all flow down from the top. In fact, directors of national laboratories had grown accustomed to considerable autonomy, and they sought to expand their laboratories' roles in addressing national issues, including environmental concerns. Thus, the AEC's expansion of this expertise also reflected competition among the commission's own laboratories for roles in environmental research programs.

For example, the director of the Argonne National Laboratory, Robert Duffield, argued that national laboratories, with their capability for big science and for linking research with practical applications, could help solve environmental and social problems.[12] The most ambitious plan, however, developed at Oak Ridge, sought to convert some national laboratories into "National Environmental Laboratories." A Senate bill based on this plan, introduced in February 1970, elicited fierce opposition from AEC commissioners and their political allies, who feared that such changes would detract from the laboratories' basic mission of nuclear energy. The Joint Committee, criticizing Weinberg's pursuit of nonnuclear research at Oak Ridge, warned of "signs that ambition to acquire new knowledge and expertise in fields outside the present competence and missions of an AEC national laboratory, in order to attain and provide wisdom which this country needs in connection with . . . its non-nuclear environmental and ecological problems, is spurring at least one laboratory to solicit activities unrelated to its atomic energy programs and for which it does not now have special competence or talents."[13] The plan for national environmental laboratories was shelved, in part because of AEC and Joint Committee objections. Not long after this episode Weinberg left his position as director of the Oak Ridge laboratory to become director of the Institute for Energy Analysis of the Oak Ridge Associated Universities.

Even so, laboratories began research in some areas, such as thermal pollution, even while the AEC was officially denying responsibility for them. Their refocusing was in tune with prevailing political opinion: Congress had criticized federal laboratories for not directing more effort toward such public problems as pollution, and it recommended that labs be given more flexibility in choosing research problems.[14] Thus, Weinberg's efforts to position Oak Ridge as a source of environmental expertise responded to these new political realities. Given his advocacy of an expanded role for national laboratories, he was himself probably partially responsible for these realities.

Even as the national laboratories differed with the AEC over environmental research, their autonomy was being displaced by more centralized control and greater emphasis on applied research. Milton Shaw, who became director of the AEC's civilian reactor development programs in 1965, felt that because reactor development was becoming commercially viable there was less need for basic research. An Oak Ridge administrator described the result: "It's no secret that there has been a trend away from science and toward more engineering work in the laboratories for a good number of years." Laboratory scientists were highly critical of this emphasis on applied research.[15]

Environmental pressures and centralized control of AEC laboratories were accompanied by a growing opinion that the AEC hindered the balanced assessment of atomic energy and other forms of energy. As a result, the 1974 Energy Reorganization Act created the Nuclear Regulatory Commission, to be responsible for licensing nuclear power stations, while other AEC operations, including research, were combined with programs from other agencies to form the Energy Research and Development Administration, itself replaced in 1977 by the Department of Energy.[16] These changes were accompanied by the expectation that the activities of the national laboratories would represent more accurately the evolving national interest in a diversity of energy technologies.

These developments would all influence Oak Ridge ecology during the 1970s. Two more immediate factors, however, were the support for ecological research provided by the International Biological Program (IBP) and the demand created by the National Environmental Policy Act (NEPA) for environmental impact assessment statements. The IBP was proposed by ecologists, approved by Congress, and funded by the National Science Foundation to provide the basis for dealing with environmental problems through an ecosystem approach, and NEPA was intended to ensure that the environmental impacts of major developments would be considered in decisions concerning their approval.

THE IBP AT OAK RIDGE

Oak Ridge ecologists became involved in the IBP in 1964, when Jerry Olson attended its first general assembly in Paris. Two years later he attended the Williamstown, Massachusetts, meeting that established the study of watershed ecosystems as the central objective of the American IBP. Reichle, with Olson's encouragement, had also by then become involved, attending in Poland a symposium on the productivity of terrestrial ecosystems. At that symposium he was elected co-chair of the Woodlands Working Group of the Terrestrial Productivity Section.[17] But like many American biologists, other Oak Ridge ecologists, including Auerbach, were initially skeptical of the IBP. By 1968, however, they had decided to participate. Given the IBP's focus on ecosystems, many thought that it could help Oak Ridge develop as a center for ecosystem studies. It would also set a precedent for funding of Oak Ridge research by the National Science Foundation, which had previously directed funds primarily to academic scientists. Auerbach saw this as a timely opportunity because several years of stagnant funding had hindered development of such projects as the Walker Branch watershed study. Funding from the NSF would help ecosystem ecology without requiring that it compete with more established AEC research projects. In 1968 Oak Ridge was chosen as headquarters of the Eastern Deciduous Forest Biome (EDFB), with Auerbach as director. Two years of planning followed before the program's official launch in 1970. Between 1970 and 1974 the IBP would provide more than $1.3 million for research at Oak Ridge, and an approximately equal amount for other EDFB activities at the laboratory.[18]

The Eastern Deciduous Forest Biome was composed of researchers at five sites: Oak Ridge, Lake Wingra (University of Wisconsin), Lake George (Rensselaer Polytechnic Institute); Coweeta Hydrologic Laboratory (University of Georgia); and Triangle Site (Duke University). Oak Ridge served as both biome headquarters and as a research site with its own program of forest ecology. Between 1970 and 1974 a large group of researchers—up to twenty-nine at one time—studied in the forests of the AEC reservation. They included scientists with a range of experience: permanent staff, visitors, and doctoral and postdoctoral students; there were ecologists with primarily biological training as well as physicists, some attracted by a presidential internship program. This diversity of experience and expertise would help shape the outcomes of the Oak Ridge research site program.

Before the program began Auerbach had determined the overall goal of the EDFB to be an understanding of the characteristics of ecosystems—their productivity, regulation, and persistence in fluctuating environments—as functions of their processes, such as primary production or decomposition.[19] Achieving this goal would require field studies and modeling on three levels. One level focused on realistic mechanistic models of ecosystem processes, such as primary production and decomposition.[20] For example, studies of primary production included various aspects of plant physiology: measurement of leaf area; comparison of leaf gas exchange with changes in solar radiation, temperature, and other meteorological variables; and the combination of these data in models relating photosynthetic activity to changes in environmental factors. In a second level of research these process models were combined with other information into models simulating the movement of carbon, water, and nutrients within an entire forest, in relation to such environmental factors as temperature and solar radiation. Results from this work were then, in the third level of research, integrated with aquatic ecosystem models into models of watersheds and other large landscape units. Overall, this hierarchy of research would be, Auerbach and his colleagues believed, a "unique contribution" of Oak Ridge to the biome program.[21]

Although this overall structure was presented as a significant innovation, much of the research consisted of detailed studies and models of specific ecosystem processes that extended lines of research already established at Oak Ridge. This reflected the goal of most biome ecologists: to study specific aspects of ecosystems using the resources available at Oak Ridge. Those already at Oak Ridge could continue research on problems for which they had data and experience. Those attracted to Oak Ridge by the EDFB, including students, could complete a project of manageable size while accumulating experience with sophisticated analytical and modeling techniques. For mathematically oriented scientists, including former physicists, the EDFB provided opportunities to gain experience in ecological modeling.[22]

As the program continued, Oak Ridge researchers began to develop synthetic ecosystem theory. In August 1972 they invited IBP ecologists from several countries to Oak Ridge to compare results and data from forest ecosystems. These data could then be used for testing hypotheses of ecosystem theory. New field studies were also intended to contribute to theory. The "Role of Consumers Project," for example, examined the mechanisms by which consumers regulate ecosystem processes, and the "Significance of Belowground Processes Project" focused on the functional role of roots and other subsurface materials.[23]

Oak Ridge researchers also developed several ecosystem models. The Terrestrial Ecosystem Energy Model (TEEM) combined models of primary production, animal consumption, and decomposition to model a terrestrial ecosystem. Using data from previous studies they tested its ability to predict ecosystem behavior. In this it had only partial success, and they cautioned that it was only an initial effort: a "starting point rather than a final answer." By the end of the EDFB such caution had been well confirmed by experience.[24] Even so, the model's developers noted that its results were useful for testing their assumptions of the significance of ecosystem processes. These assumptions might, they expected, "form the grounds of spirited scientific debate."[25] Thus, ecosystem models, they argued, had a role to play in debating and building ecological theory.

While the largest component of EDFB activity at Oak Ridge was research and modeling of forest ecosystems, the management by Auerbach and his colleagues of the entire biome program was also significant. Their strategies provide clues to their views on the future of ecological research and on its potential role in environmental management.

Auerbach intended for the EDFB to enhance the reputation of Oak Ridge for large projects combining research and modeling in systematic study of an ecosystem. Accordingly, Auerbach assembled a small team of managers to facilitate the biome program's evolution into an integrated effort. They used several methods: newsletters, reports, meetings, biome-wide studies, and systems analysis techniques, such as Program Evaluation and Review Technique charts, that showed how each project contributed toward the overall objectives. The most important method of integration was to be the "systems approach": the use of models and related techniques that had proved useful in studies of other complex systems. An ecosystem model would serve as a framework for submodels and facilitate planning and data gathering. Empirical results would contribute to modification of the models, analysis of which would then suggest further research. Other ecologists, they hoped, would emulate this combination of models and empirical research. It was, they declared, an "essential part of the evolution of future ecological research." At the conclusion of the project the models would provide a synthesis of understanding of the ecosystem or of particular aspects of it.[26] Models would therefore both be a tool and constitute a major product of the research.

These management strategies would themselves be objects of research. "A major challenge will be the creation of an effective functional administrative organization in the biome," Auerbach and his colleagues noted in 1970. "The organization outlined . . . will provide a starting framework for meeting this challenge. Only through operating experience, however, can the challenge be tested."[27] In effect, management itself was an experiment: just as results of field research would suggest changes in models, the results of the management process would suggest ways to improve the process itself. For Auerbach and his management colleagues, therefore, the objectives of the EDFB extended beyond understanding ecosystems to perfecting research strategies for complex systems, including ecosystems.

The EDFB's status as a large integrated project therefore reinforced Auerbach's persistent argument that ecosystem research could be big science. The EDFB, he claimed in 1973, had demonstrated to both the AEC and the scientific community the value of the "ORNL style" of research to ecology.[28] In part, this linking of EDFB research to the Oak Ridge "style" continued his 1960s strategy of building his program around large projects to show that ecology belonged at Oak Ridge. Only the source of funding, and the attention devoted to organization, had changed.

Integrated research in the ORNL style was also, biome researchers stressed, better able to solve environmental problems than was small-scale individual research. "A piece-meal approach which studies individual processes by traditional methodology," Robert O'Neill argued in 1971, "cannot be depended on to provide an adequate predictive capability within a useful time-scale." The use of systems analysis and modeling to study complex systems could, Auerbach and Reichle believed, provide a new approach to such environmental management tasks as impact assessment and site selection for generating stations and other developments. "[We] are developing insights into research management," Auerbach and biome deputy director Robert Burgess told a congress of biologists in 1971, "that can only, in the long run, provide mechanisms that have the potential for lasting benefit to society as we struggle with environmental problems." For Auerbach, demonstrating this could help ensure a central role for Oak Ridge in the research programs expected to follow the IBP. "If the pattern of team co-operation is successful," he predicted, "it could well be the fount for innovative approaches to environmental research in subsequent years."[29]

Weinberg, the director of the Oak Ridge laboratory, shared this perspective. He had for several years argued the need for large-scale integrated research to solve complex national problems. He supported involvement in the IBP because he agreed with Auerbach that, as integrated research, it was highly appropriate to Oak Ridge: "The integrated ecosystem studies . . . of the U.S. plan represent an extension of an effort which was initiated by AEC environmental programs and endorsed by our Review Committees. . . . Many of the individuals who are best prepared to fill leadership positions in such research received much of their stimulation or experience in the AEC laboratories."[30]

More generally, by the late 1960s many scientists and government officials in the United States believed that societal problems, including environmental degradation, were caused by fragmented, uncoordinated decisions and policies. Decisionmakers too often simply balanced special interest groups and neglected the broader public interest. Changes in society were also significant: rising affluence had led Americans to aspire to more than strictly economic goals; to seek not just the necessities of life but those amenities, such as clean air and water, provided only by a protected environment.[31] Such aspirations helped justify, for many, a positive role for government, in meeting societal priorities that could not be satisfied by the marketplace or by the balancing of special interests. Government, instead, could develop stronger roles in resource-allocation and regulation; these roles would rely on comprehensive, rational decisionmaking, using the insights of many disciplines.[32] As Robert Lilienfeld has argued and Chunglin Kwa has noted with respect to ecosystem research, this view may have been reinforced by the notion, widespread among "scientific and technocratic elites," that society and nature form closed systems that could achieve stable, optimum operation through rational expert analysis and management. The widespread adoption of systems analysis and other management strategies within the American government reflected the perceived need for comprehensive, rational decisionmaking to deal with complex problems.[33]

To some extent, Auerbach and his biome colleagues employed the rhetoric characteristic of this view, presenting as a chief objective of their program the demonstration that, just as systems analysis had proved useful in the study and management of other complex systems, so could its application to ecosystems help provide a rational approach to optimum use of the environment. Replacing "simplistic, narrow approaches," systems analysis would permit scientists to study the full complexity of ecosystems by linking the research of every discipline relevant to understanding ecosystems.[34] The use of systems analysis techniques in the EDFB showed that greater attention was being paid to its organization and management than had been present in previous ecosystem research projects at Oak Ridge, such as the cesium forest, or at Merlewood. The difference was, in part, one of scale: the EDFB had to be organized more carefully simply to ensure that its huge investment of time and money would generate worthwhile results. The EDFB's greater emphasis on synthetic eco-system theory also demanded that contributing studies' results be available when required, thereby requiring their careful organization. Finally, the EDFB was intended, as earlier projects were not, to demonstrate how ecologists could be organized to tackle large and complex problems requiring comprehensive approaches.

Biome research, including the work at the Oak Ridge site, provided the largest demonstration of the application of systems analysis to ecosystems. To buttress their argument that systems analysis had practical utility, biome researchers, led by O'Neill, applied models to various environmental problems. In one study O'Neill and O. W. Burke used techniques predating the IBP to simulate the movement of DDT and DDE through the food chain to humans. This was at the request of the DDT Advisory Committee of the Environmental Protection Agency. The committee's chair, Orie Loucks of the University of Wisconsin, had previously used systems methods to integrate scientific evidence in the courtroom. EDFB ecologists also collaborated in studies of changing land-use patterns in eastern Tennessee, the regional effects of a power plant, the impact of supersonic transports on forests, and other projects.[35]

Each of these projects attempted to demonstrate that simulation models could help in understanding ecosystems or in predicting their future states. It is significant, however, that ecologists doing these projects saw them as trial runs performed with limited resources, and they usually concluded that more definite results required further study. Further study usually did not occur, however. In part, this was because, as O'Neill and other ecologists had explained, development of ecosystem models able to provide useful predictions proved more difficult than originally envisaged. Ecologists elsewhere, however, did pursue systems analysis to the extent of deriving lessons for human society from the characteristics of ecosystems.[36] Evidently Oak Ridge ecologists did not consider systems analysis of society within its natural environment to be a chief aim of their study.

In fact, ecosystem ecology at Oak Ridge, during the biome program and after its conclusion in 1974, remained focused on the study of natural ecosystems in equilibrium. During the 1970s O'Neill, Reichle, and other Oak Ridge ecologists defined homeostatic capability—the maintenance of stability in fluctuating environments—as the central theoretical problem of ecosystems.[37] Because the stability of an ecosystem could not, they argued, be predicted from individual study of its components, ecosystems themselves were the appropriate unit of study. Seeking the mechanisms of homeostasis, they identified three functional components common to all ecosystems: an autotrophic base, to capture energy; heterotrophic consumers, to regulate process rates; and a pool of organic matter, to serve as a reservoir of stored nutrients. Each component, they argued, contributed to ecosystem homeostasis.[38]

Oak Ridge ecologists also argued that ecosystem theory was relevant to practical environmental problems, as it could, for example, assist in detection of damage to the mechanisms that ensure stability. The state of ecosystem homeostasis gave, they stressed, a better indication of the condition of the natural world than did those aspects—abundance of game and species diversity, for example—most valued by society.[39] Particularly noteworthy was the argument that ecosystem theory could provide a long-term basis to environmental quality by anticipating problems before they became widely recognized. This, ecologists insisted, complemented the tendency of environmental management to focus on short-term, immediate concerns. Study of forest productivity and carbon cycling, for example, could help predict the eventual problems that could arise from changes in atmospheric carbon dioxide.[40] Justifying ecosystem research in terms of distant problems, of course, meant that it could be described as relevant to environmental problems, without having to provide immediate solutions. Oak Ridge ecologists stressed that immediate solutions should not be expected. Although "clues for intelligent ecosystem management lie in an understanding of dynamic ecosystem processes," Reichle and a colleague noted that a "wide range of ecological questions must be answered before complex environmental problems can be resolved."[41]

This stressing of the role of ecosystem ecology in anticipating environmental problems reflected a change in how Oak Ridge ecosystem ecologists perceived their role within society. At the start of the IBP they had argued that systems analysis of ecosystems could deal directly with the most complex problems engendered by society's interaction with its natural environment. This argument was consistent with the widespread belief that systems analysis promised optimal management of social and natural systems. By the close of the 1970s, application of this novel method was no longer seen by Oak Ridge ecologists as the major innovation of ecosystem ecology; instead, theory would be the arena of innovation. The role, furthermore, of ecosystem theory in environmental management would be more indirect than that once envisaged for systems analysis of ecosystems. Theory could better anticipate problems than solve actual problems. In effect, ecologists had returned to a traditional justification of basic science as able to provide theoretical understanding of nature, of eventual utility, but not necessarily immediate practical significance.

Attention to these perspectives on the relation of ecosystem ecology to environmental problems, expressed primarily by the EDFB's leadership, should not obscure the striking complexity of the overall program. A variety of motivations and priorities shaped it: Auerbach's concern for research management; ecologists' interest in ecosystem theory; other scientists' and students' emphasis on specific ecosystem processes; modelers' focus on modeling complex systems. To a large extent these priorities influenced this decentralized program, in which scientists were given wide, if not unbounded, latitude to design their own research strategy.[42] A similarly wide variety of products was generated by the EDFB: ecosystem and process models, some of which would be used in later research; ecosystem theory; experience in practical application of ecosystem models, and awareness of their limitations; and perhaps most important, trained students who had some of their first research experience at Oak Ridge. The EDFB, in short, was not a unitary phenomenon but a complex, heterogeneous combination of the interests and priorities of its participants.[43] In interpreting the IBP, historians have tended to obscure this complexity by focusing on the programmatic assertions of its leadership or the ideological implications of the systems perspective portrayed as underlying the program. Or, they have most often focused on the Grasslands Biome project, the part of the IBP that most clearly epitomized the hierarchical, "top-down" form of research organization.[44] As Oak Ridge research shows, however, the IBP was not strictly a "top-down" research program. The work of each ecologist or modeler reflected that individual's own ambitions and experience as much as it did the plans of biome directors.

This approach to management of research was one element of the ultimate impact of the EDFB on Oak Ridge ecology. It engendered a shift away from a research style based on the institutional and intellectual environment of the national laboratory toward one corresponding more closely to that of the academic community. The decentralized organization of the program was one manifestation of this. Oak Ridge ecologists' interest in theory was another. Still another was the transformation of the methods of Oak Ridge ecosystem research. The dominant technique of the 1960s—of tracing the movement of radionuclides and deriving from this the movement of energy and nutrients within ecosystems—became in the 1970s only one of several techniques used. Instead, new methods, as well as theoretical insights, were absorbed from the academic research community. Study of cycling and loss of nutrients from ecosystems, for example, was stimulated by the results of the Hubbard Brook Ecosystem Study.[45]

The largest factor behind the convergence of Oak Ridge ecology with academic research was the National Science Foundation and its peer-review system of funding allocation. The biome program ended in 1974, succeeded by a period of synthesis. Oak Ridge ecologists also pursued several smaller projects, some continuing work begun during the IBP. These included studies of primary production, stream ecology, the role of consumers in ecosystems, the global carbon cycle, and interactions between aquatic and terrestrial ecosystems.[46] In most cases these projects were led by scientists who had participated in the EDFB. They were also funded by the NSF. With the end of the foundation's support for the IBP in 1974, Oak Ridge ecologists competed directly with academic ecologists for research funding. And their success within the NSF peer-review system depended in part on their reputations among academic researchers. Thus, the irony of the IBP at Oak Ridge: brought to the national laboratory to further its distinctive big science approach to research, it actually shifted its approach closer to that of university ecologists. This included a shift in how Oak Ridge ecologists perceived their role in relation to environmental problems: from tackling such problems through large-scale, comprehensive research to anticipation of emerging problems and development of a theoretical basis for understanding them.

Even as ecosystem ecologists at Oak Ridge continued their exploration of ecosystem theory, and while doing so, redefined their role in environmental affairs, other events within American society generated new demands for expertise in environmental science. These demands would constitute a second major factor shaping the role of Oak Ridge ecologists in environmental affairs.

ENVIRONMENTAL IMPACT ASSESSMENT

In 1970 the National Environmental Policy Act came into effect. One of its most significant requirements was that each federal agency proposing a major project prepare a statement assessing its environmental impacts and explaining how adverse impacts could be avoided or mitigated. In response, the AEC began preparing environmental impact statements for new nuclear power stations. Relatively cursory and limited to radiological hazards, these statements epitomized the commission's reluctance to accept responsibility for the environmental risks of nuclear power. In July 1971, however, a Maryland court ruled that the impact statement prepared for a nuclear power station at Calvert Cliffs was inadequate, and it ordered the AEC to prepare more thorough assessments.[47]

This put the AEC into a state of near-crisis. Because new plants could not be licensed until they fulfilled NEPA requirements, the decision threatened to halt, if only temporarily, expansion of nuclear power. AEC officials accordingly ordered Oak Ridge and two other laboratories to assume responsibility for preparing impact assessment statements. Oak Ridge management, in turn, ordered Auerbach and his staff to take on what became an enormous task. After July 1971 they were, at any one time, involved in preparing between ten and fifteen impact statements. By September 1973 they had helped prepare forty-one statements, primarily for nuclear power plants.[48]

Auerbach welcomed this new responsibility. More "fundamental progress in ecology," he told ecologists, "seems likely if the relevance of our discipline is thoroughly understood throughout government, and not just by the popular press."[49] With the media-christened "age of ecology" in full swing, Auerbach evidently hoped that impact assessment would help demonstrate that ecology was not just a popular movement but a discipline of practical utility. But this new task was also disruptive. To cope with the workload Auerbach reorganized the division, hired new scientists, and reassigned others to impact assessment. Several projects were curtailed so that researchers could work (some reluctantly) on impact statements.[50] Auerbach, however, excluded EDFB scientists from this reassignment because that project was already under way and was funded by the NSF, not the AEC.

Beyond increasing the workload, impact assessment imposed novel demands on Oak Ridge ecologists. To understand why, it is necessary to examine how impact assessment statements fit into decisionmaking. These statements were prepared for the AEC's regulatory branch until 1974, when they were prepared for the Nuclear Regulatory Commission. They became an essential element of the power plant licensing process. Decisions on plant licensing were delegated to Atomic Safety and Licensing Boards (ASLB). These boards, as Oak Ridge ecologists learned, based their decisions on a few key issues: Would there be an adverse impact? If so, would it be outweighed by the project's benefits? If the project was otherwise acceptable, were cost-effective modifications available to reduce its environmental impact? In effect, the process consisted of balancing the socioeconomic and environmental costs and benefits of the power plants. In making its decision, an ASLB considered impact statements, other engineering and environmental data, and the testimony of witnesses.[51] Ecologists also had to consider the legal requirements affecting impact assessment. Federal Water Pollution Control Act Amendments, for example, required fish population and aquatic biology studies near thermal generating stations. In addition, AEC guidelines established which species were "important," and therefore merited consideration.[52]

A complicating factor in this process was its adversarial nature. Licensing commonly involved three parties: the applicant (usually a utility); the regulatory agency (for whom Oak Ridge ecologists prepared their studies); and intervenors (including private citizens and environmental groups). Each party could hire its own scientists, who often disagreed as to the predicted environmental impacts, costs of their mitigation, or the project's benefits. As each sought to persuade the ASLB, legal counsel would examine the opposing side, seeking weaknesses and forcing it to justify its predictions. Impact proceedings were also open to public scrutiny, and protests were often part of the process.

These circumstances all affected how Oak Ridge ecologists prepared for the "rigors of defending their predictions or conclusions in public forums or in adjudicatory hearings."[53] Assessments tended to focus on impacts that could be expressed in monetary terms or affected aspects of the environment of special interest to the public. To be defensible in the courtroom, research results required simplification, a focus on the strongest conclusions, and neglect of important but uncertain issues. Most important, the impact statement, to be credible, had to be seen as based on objective scientific data, independent of unwarranted assumptions, value judgments, and partisan advocacy.[54] Witnesses who made statements considered "nonscientific" risked losing credibility. Oak Ridge ecologists, accordingly, concluded that they would be most credible if they relied strictly on quantitative evidence, such as simulation models. As Auerbach noted, "It is not easy to separate professional judgment from personal value judgment; this is probably one of the most challenging obstacles we as ecologists face and is one of the strongest reasons for the further development of the quantitative aspects of our science." Although they gave value judgments when requested, ecologists insisted that these were separate from the impact statements, which should be considered to be based strictly on scientific evidence.[55]

These conditions contrasted with those once prevalent at Oak Ridge and at the Nature Conservancy. Because of the low level of concern about the impact of nuclear energy and its insulation from political controversy, Oak Ridge ecologists in the 1960s did not have to present results (concerning, for example, the potential impact of Plowshares projects) within adversarial processes. And, as we have seen, the application of ecological expertise to conservation by the conservancy usually occurred through cooperation and consultation, not confrontation between adversaries.

The demand generated by impact assessments for scientific evidence reflected the insistence within American regulatory proceedings on evidence that could be considered factual and value-free and would thereby impart an aura of objectivity to the resulting decision. As one observer has noted, "Both decision-makers and advocates . . . have strong incentives to claim that their actions and arguments are grounded in knowledge, not interest or values, and can often thereby muster support. . . . The pretense of knowledge yields power."[56] Within legal and quasi-legal settings (such as ASLB hearings), this demand for objective facts as the source of authority for decisions has most often led to a reliance on scientific evidence.[57] That "scientific" evidence, in turn, has been generally defined within these settings as data obtained through quantitative methods that seemingly eschew value judgments.

The effect of these demands on the work of Oak Ridge ecologists can be understood by examining first their efforts to assess the risks of exposing humans to radioactivity, and second, their participation in an actual controversy over the impact of several nuclear power plants on the Hudson River north of New York City.

RADIATION HAZARDS

In the early 1970s concern over emissions of radioactive materials from nuclear plants focused not so much on their effect on the environment as on their potential hazard to humans.[58] To predict radiation exposure from plant operations involved understanding both the pathways by which radioactivity released into the environment could come into contact with humans, and knowledge of human habits and biology that influence radiation dose. It therefore required the expertise of both ecologists and health physicists. They were combined within a new Environmental Hazards Evaluation program, which in May 1972 was transferred from the Health Physics Division to the Environmental Sciences Division.

Before 1971 Oak Ridge ecologists had made only a few attempts (most notably in the assessment of the proposed Panama Canal project), to predict radiation dose to humans from environmental contamination. In coping with the demand for predictions created by the National Environmental Policy Act, ecologists argued that the methods they had used in these attempts—systems analysis and simulation modeling of the movement of radionuclides along food chains to humans—would be essential. These techniques, Auerbach expected, "could be used as a significant tool in public relations to add objectivity to environmental assessments, and to help formulate new guides for determining acceptable releases of radioactivity into the environment." The experience they had gained by 1971 was only a beginning, however. As they often noted, their model, limited to the agricultural food chains leading to humans, could not provide reliable predictions. In particular, lack of data and of understanding of environmental transport mechanisms forced it to rely on assumptions (for example, that fallout was the only source of radionuclides) too simplistic to be credible in a licensing process.[59]

One immediate need was for more rigorous, quantitative data on the movement and accumulation of radionuclides within organisms. While scattered information was available, it fell short of what was required. As they noted in 1971, "It is one matter to know that food chains are important but quite a different matter to quantify these complex chains for natural and agricultural ecosystems. . . . We have neither sufficiently detailed nor widely representative radioecological data with which to work."[60] A second need was for the basic food chain model to be combined with information on the movement of radionuclides through the abiotic components of the environment: their aerial dispersion and movement through groundwater and sediment, for example. Ecologists had to collaborate with meteorologists, hydrologists, and geologists in linking models for these components with their own model. A third need was to link modeling of the environmental dynamics of radionuclides with the internal and external dosimetry models developed by health physicists.

As quickly as the demands of impact assessment permitted, researchers of the Environmental Hazards Studies Section, led by Stephen Kaye, set about refining the model, replacing assumptions with more rigorously modeled functions, linking it to submodels, and improving its realism and thereby the credibility of its predictions. Their goal was to develop one standard method for prediction of radiation exposure to humans. By 1973 they had such a method, called CUEX (cumulative exposure index), and had begun to apply it to impact statements. CUEX was designed both to predict the radiation dosage to individuals and populations resulting from the release of any radionuclide and to compare this to the maximum annual dosages permitted by radiation protection regulations. It consisted of a series of models simulating the movement and accumulation of radionuclides within various parts of the total environment, including the atmosphere, terrestrial and aquatic food chains, hydrological transport mechanisms, and sediment. Each radionuclide likely to be released by nuclear facilities was considered separately; by 1976 they numbered 185.[61]

While using CUEX in impact assessments Kaye and his colleagues continued to refine it, seeking more credible predictions and fewer assumptions. In 1976, for example, one project evaluated the reduction in predicted radiation dose that could be expected when the fact that humans spend most of their time indoors was considered.[62] This effort to refine CUEX differed in several respects from the studies of radionuclide movement through ecosystems characteristic of Oak Ridge ecology during the 1960s. They sought not to develop implications for energy flow and material cycling from the movement of a single biologically and experimentally tractable radionuclide such as cesium 137 but greater precision in predicting the movement of all radionuclides likely to be released by energy facilities. In effect, interest had shifted from the cyclic movement of materials to the linear flow of radionuclides from a nuclear plant, through the environment, and to humans.

POWER STATIONS ON THE HUDSON

The controversy concerning nuclear power on the Hudson River began in 1964, when a proposal for a pumped-storage reservoir on the river elicited fierce opposition.[63] By the early 1970s attention had shifted to the Indian Point nuclear power plant. Although the first reactor at this plant had already been constructed and licensed according to earlier, strictly radiological criteria, the nonnuclear impact of its second unit now had to satisfy NEPA requirements before an operating license could be issued. Oak Ridge ecologists became involved in 1972, when they were asked by the AEC to serve as technical experts.

They had, they found, entered a dispute between sharply defined adversaries. The AEC was seeking to evaluate the impact of the station in order to fulfill its NEPA obligations. Local citizens and environmental groups were protesting the effects of this and other generating facilities on the Hudson River. Consolidated Edison wanted to prove that their plant had a minimal impact on the river. Otherwise, they might be required to install cooling towers and other facilities to reduce demand for cooling water, at a cost of several hundred million dollars. To support their arguments they and other utilities provided up to $5 million per year for huge research programs on the biology of the Hudson River.[64]

A focus of the controversy was the striped bass. The basis for a large sport and commercial fishery, these fish could be decimated by power plant cooling systems, through impingement (being trapped against screens installed on intake pipes) or entrainment of eggs and larvae (being transported through the cooling system of the plant to be killed by mechanical and thermal stresses). These impacts had been of concern since the initial controversy over the storage reservoir, and in the mid-1960s, study of the project's potential impact had focused on this species.[65]

This study of one species may appear uncharacteristic of ecologists located at a laboratory known for ecosystem research. The ecologists explained, however, that the striped bass, as a representative anadromous species and one of the top carnivores in the Hudson River Estuary, was significant ecologically. In addition, the utilities' research programs were already providing information on its distribution and abundance. A more convincing assessment could be done, they knew, on the basis of those aspects of the environment already well studied.[66]

Sigurd Christensen and other Oak Ridge ecologists also argued that a focus on a single fish species rather than on an entire ecosystem was better suited to the constraints imposed by the impact assessment process. The cost and time involved and the uncertainty of prediction made an ecosystem approach impractical.[67] Such an approach, emphasizing such functional properties as productivity or nutrient cycling, was also not as relevant to public concerns. While power plant cooling systems could affect fish populations they may not have an effect on ecosystem properties. Even if an effect was found it would not contribute to resolving the dispute over impacts on those aspects of the river environment of value to the public. It is evident, therefore, that the choice of Oak Ridge ecologists to study the striped bass was actually made at the start of the controversy in the 1960s, several years before they actually became involved.

In their use of simulation models, however, their approach was characteristic of Oak Ridge. They had realized that their assessment would be considered most objective and creditable if it relied on quantitative evidence, including simulation models.[68] Their first, relatively simple model was developed in 1972 by Philip Goodyear and Charles Hall. It simulated the movement of fish eggs, larvae, and juvenile fish in response to water flow and temperature.[69] In 1973, on the basis of this model and models prepared by the utility's consultants, the ASLB ruled that neither the utility's prediction of minimal effect on fish populations, nor the AEC's prediction of substantial depletion, were conclusive. But the board also concluded that it would be prudent to accept the AEC's predictions, and it ordered the utility to reduce its need for river water by installing a cooling tower. The following year Consolidated Edison successfully appealed this decision, reopening the dispute.[70]

For the next several years both sides developed ever more sophisticated mechanistic simulation models, all aimed at predicting the long-term impact of impingement and entrainment on Hudson River striped bass populations. Oak Ridge ecologists developed two such models, designed to work in combination. The young-of-the-year population model predicted how many fewer striped bass would survive their first year (when they were most vulnerable to impingement and entrainment) because of the presence of the power plants. This prediction was then used in a life-cycle population model to predict the long-term impact of increased first-year mortality.[71]

These highly ambitious models were based not on actual population data but on what was known of the relation of environmental variables to the biology of the fish populations. The young-of-the-year model, for example, related physical variables, such as temperature and salinity, to spawning and growth rates, maximum swimming speeds, and other biological variables.[72] A second ambitious goal, as I have noted, was prediction of not just the year-to-year mortality of the fish but the long-term viability of Hudson River fish populations. In making such a prediction the ecologists had to form assumptions about the behavior of fish populations that, while justifiable in terms of their professional experience, could not yet be fully verified. In particular, compensation (the tendency of populations to increase their growth rate when experiencing increased mortality) was difficult to quantify. This remained a significant problem, despite extensive research at Oak Ridge and elsewhere. The task of prediction was further complicated when the assessment was widened to include six power plants along the valley. In response, Oak Ridge ecologists extended the scope of their assessment to the entire lower Hudson River.[73]

Even as these models were refined and became increasingly complex, they failed to provide a basis for ending the dispute. The central difficulty was that the models used by both sides generated predictions that were both highly uncertain and sensitive to assumptions about imperfectly understood phenomena, such as compensation. Utility scientists tended to make assumptions that minimized impacts on fish populations, and Oak Ridge scientists tended to predict larger impacts.[74] With scientists unable to agree on their predictions, the regulatory agency was unable to choose one over the other. The role of science in such proceedings, as I have noted, was to provide an objective basis for decisionmaking. But as it became obvious that science was unable to provide any such basis, it also became clear that no decision could be reached.

By 1977, when hearings reconvened before the Environmental Protection Agency, both sides had developed simpler approaches. Instead of simulation models based on biological and physical processes, data describing distribution and abundance of striped bass and other species were used to predict entrainment and impingement directly. Oak Ridge ecologists put aside their complex simulation models and developed an Empirical Transport Model. Its advantage was that it did not depend on assumptions (particularly concerning the relation of hydrodynamic processes to fish distribution) that could not be corroborated quantitatively. It was also easier to explain to nonscientists.[75] But it was less ambitious, predicting only the abundance of individual year classes, and not the long-term viability of fish populations.

Unlike the simulation models, however, this model could predict fish mortality in specific circumstances: if, for example, cooling towers were installed, or plant operating schedules were modified. Such predictions formed a basis for negotiations that began in August 1979 between the EPA, utilities, New York state agencies, and environmental groups. On December 19, 1980, the groups reached a compromise settlement in which the utilities agreed to implement measures that would reduce, if not eliminate, fish mortality, and were less expensive than cooling towers.[76]

In this episode the constraints of the impact assessment process overruled ecologists' initial perceptions of the appropriate approach. Oak Ridge ecologists had presented ecosystem ecology as best able to assess the impact of human activities on the environment. In an actual impact assessment, however, the focus was not on the ecosystem but on one part of it—the striped bass—relevant to immediate social and economic concerns and best suited to quantitative predictions. Then, after several years of effort, it became apparent that another distinctive technique of Oak Ridge ecologists, simulation modeling of the river environment, was less suitable than other methods relying on local empirical evidence. To fulfill the demands of the process they had to adhere to its constraints, reworking their approach so as to contribute to an agreement between adversaries.

In these proceedings can be observed both the evolution of a new context for science and an effort by scientists to comprehend their role within it. They entered the hearings believing that by predicting the long-term impact of entrainment and impingement on fish populations, and by quantifying the significance of key ecological processes, including compensation, they could provide the basis for agreement while advancing scientific understanding of fish population dynamics. But none of these scientific objectives was attained before the settlement was reached. What the scientists did provide were empirical models able to demonstrate how specific power plant modifications would affect fish populations. Such models provided a basis for negotiation between government agencies seeking to minimize impacts on fish populations and a utility aiming to meet regulatory requirements with the least expense. Like many environmental controversies, this impact assessment process was not primarily a technical issue with an objective solution obtainable through scientific research but rather a political issue consisting of a conflict between opposing values and priorities. Research provided the basis for negotiation and compromise through which agreement was reached but could not itself replace these essentially political processes.

ECOSYSTEM AND ENVIRONMENTAL RESEARCH

Environmental impact assessment was only one, albeit the largest, of the new tasks imposed on the Oak Ridge ecology program during the 1970s. Environmental research programs proliferated as government and business sought to comply with or evade environmental regulations. Oak Ridge scientists participated in many of these programs on such topics as the environmental behavior of toxic substances and the environmental aspects of synthetic fossil fuels, plutonium, radioactive waste disposal, and coal combustion. These projects reflected the broadening of its research into nonnuclear areas, as a result of the conversion of the AEC into the Energy Research and Development Administration in 1974, which then became the Department of Energy (DOE) in 1977. The environmental research program grew continually throughout the 1970s: by the end of the decade the Environmental Sciences Division employed nearly 200 scientists and other staff. In 1978 the division gained a new building to house its rapidly expanding activities.[77]

Accompanying this wide range of research was a broadening of the Environmental Sciences Division's funding support. In the 1960s ecological research at Oak Ridge had been a single, relatively unified research program, supported primarily by the AEC's Division of Biology and Medicine. In the 1970s it became the sum of a range of more narrowly defined research projects tied to specific environmental issues and sponsored by numerous agencies. By 1979 the ESD was receiving funding from three sections of the DOE (Environment, Energy Technology, and Resources Applications), the Nuclear Regulatory Commission, the EPA, the NSF, the Council for Environmental Quality, the Department of the Interior, the Electric Power Research Institute, and the Tennessee Valley Authority.

Some of these new tasks, like impact assessment, had been assigned to the division. In other instances Auerbach and his colleagues identified potential research areas and then marshaled support from federal and state agencies to study them. In 1971 a program of "applications research" was initiated. Its purpose was to begin research in areas for which the ESD was not yet directly responsible but could be in the future if it were able to demonstrate the relevant expertise.[78] Overall, therefore, they sought both to respond to the immediate needs of sponsoring agencies and to expand their own activities by anticipating requirements in environmental research.

The resources directed into Oak Ridge environmental research reflected the proliferation of environmental research that occurred during the 1970s, as regulatory agencies generated demands for scientific knowledge about the environment. Their demand for environmental science was sharpened by the political context of these agencies. Administrative agencies, such as the EPA, are often thought of as politically neutral, with a role limited to the efficient implementation of policy. In fact, they exercise considerable discretion in interpreting imprecise statutory mandates, balancing conflicting values, or determining regulatory priorities. Because of this discretionary authority, conflict often arises between competing interests as they struggle to ensure that their views are reflected in agency decisions. Often the role of the agency will be to broker these interests and distribute the costs and benefits of its decisions among those affected. Such a role derives some legitimacy from an underlying assumption of American government that good public policy is most likely to arise out of the structured clash of interests. It is further reinforced by the dependence of American administrative government on judicial or quasi-judicial proceedings, which often exacerbate conflict by emphasizing the adversarial status of competing interests. However, even as agencies exercise their discretion and maneuver among competing interests they seek to ground their decisions in scientific methods and knowledge. Demonstrating a scientific basis for their decisions supports agencies' claims that their decision-making process is rational and objective and therefore cannot as readily be challenged.[79]

This can be illustrated by a brief review of the Hudson River dispute. It originated in a conflict between environmentalists and utilities. Two agencies—first the Atomic Safety and Licensing Board, and then the EPA—attempted to develop through environmental impact assessment a scientific basis for resolving these competing interests. In effect, a clash of values and interests—between those wishing to preserve fish populations and those aiming to minimize the cost of power generation facilities—became redefined as a technical dispute. As such, it would supposedly be amenable to rational, objective solution. Such a belief helped generate the great demand for environmental research in the 1970s.[80] For Oak Ridge ecologists and environmental scientists the chief implications of this belief, as I have noted, were a broadening of research support and transfer of much of the direction of this research, from the laboratory itself to sponsoring agencies, as research funds became more closely tied to immediate regulatory and policy responsibilities.

This portrayal of environmental affairs as a structured clash of interests contrasts with the view that the environment should be subject to rational, expert management in which sectional interests and fragmentary perspectives would be secondary to the advancement by government of a broader public interest, using an integrated systems perspective. This alternative approach became increasingly untenable in the early 1970s because of at least three factors: the loss of a consensus within American society that urgent action was required in environmental protection; a decline in public trust in the ability of government to serve society's best interests; and a growing resistance to active intervention by government in the economy and society. As this consensus and trust were lost and as government sought a less active role, it became unlikely that interest groups, such as utilities and environmental groups, would defer to experts promising rational, comprehensive management. Instead, they sought, with the assistance of statutes such as NEPA, to participate in policy and regulatory decisions.[81]

As I have noted, however, in the 1960s ecosystem ecology was envisaged as most useful within a comprehensive systems approach to environmental management. As this approach became less viable within American society and its government, ecosystem ecologists were obliged to revise the practical justification of their research. As I noted above, at Oak Ridge this revision was, in part, the consequence of experience gained in the IBP. However, I suggest that underlying the transformation in the perceived relevance of ecosystem ecology—from providing a basis for environmental management (as envisaged at the start of the IBP) to anticipating more distant environmental problems—were deeper shifts in society's views of the urgency of environmental problems and of the appropriate role of government in society.

During the 1970s, ecosystem ecologists, as I have argued, developed a greater concern for theoretical issues in ecology, such as the mechanisms of ecosystem stability. As they developed stronger ties with academic ecologists, they increasingly contributed to debates within this community. Because such debates were, however, usually remote from the priorities of agencies funding most Oak Ridge research, ecosystem ecologists in their work came to have less contact with specific environmental policies and regulations.[82] This trend was encouraged by organizational changes. By the mid-1970s some ecologists had come to focus entirely on basic ecosystem ecology and the more recently acquired environmental research responsibilities were assumed by scientists specifically hired for such tasks. In 1974, for example, staff responsible for environmental impact assessment were placed in a separate unit within the division.

The impetus to relate ecosystem ecology to immediate environmental controversies also diminished as its funding shifted to the NSF. While the foundation resisted pressures to support research on specific environmental issues, the AEC, its successors, and other government agencies became increasingly reluctant to fund ecosystem research not directly tied to their regulatory or policy responsibilities.[83] In addition, the increasing sophistication of both ecosystem ecology and applied environmental research, the first engendered in part by the methods and theoretical constructs stimulated by the IBP, the second by the requirements of impact assessment and similar processes, demanded longer and more specialized training. This made it more difficult for ecologists to be active in both areas, creating a sense in each of a separate professional identity. Those active in research fields directly applicable to environmental regulations and policy became known as environmental scientists, and they resisted being identified as ecologists.

This separation of ecosystem ecologists from those studying actual environmental problems was reflected in several aspects of Oak Ridge research. Separate data bases were developed for each group. At the start of the EDFB program, for example, a database was established that was hoped to be of use in conducting environmental impact assessments. It has, however, been used mostly by those ecologists who contributed to it, while impact assessment researchers have developed their own data bases. Divergent perspectives on modeling also emerged. While ecosystem models were not capable of precise prediction, ecologists still considered them useful as research tools. One ecologist (not at Oak Ridge) summed up this view: "As long as your model output is wrong you continue to learn something. Once it fits the data you do not know what to believe. It is then predictive but it probably will provide no further insights into the dynamics of the system."[84] Models were valued not for predictive accuracy, but for their ability to guide further research and to suggest links between ecosystem phenomena. Those involved in actual environmental problems had a very different view. As the experience of impact assessment researchers showed, the function of models in regulatory and legal proceedings was to provide predictions that would be considered objective and accurate, thereby fulfilling the role assigned to scientific results within adversarial proceedings.

OAK RIDGE, AND THE ROLES OF EXPERTS

In the 1960s Auerbach and his colleagues planned a new building for ecology. The building would be round: a highly distinctive design within the laboratory. This design would also encourage interaction among researchers—architects had found that individuals meet more often in buildings with fewer corners. Most important, the building would integrate the components of ecosystem research: "We had designed the round building to serve as the functional center for a network of field operations, in that samples and field collections flow into the building, flow to the laboratories, move from the labs in the form of data down to the computers, and then out in the form of publications."[85] Designed to facilitate ecosystem research, it was, by analogy—functional components, integrated by the flow of the materials and information—itself an ecosystem.

The round building was never built. The immediate problem was cost: in the 1970s the AEC could accept only a more economical rectangular building.[86] But changes in architectural plans were at least symbolic of a shift in both ecological research at Oak Ridge and its surrounding social and institutional context. In the 1960s, ecosystems were the focus of ecological research at Oak Ridge. In the 1970s, they became only one, and not the largest, of several research areas within the Environmental Sciences Division.

The early 1970s marked a watershed for Oak Ridge ecology. The IBP brought new funding for ecosystem research. NEPA and other environmental policy initiatives provided new research and regulatory responsibilities. With the demise of the AEC the scope of Oak Ridge activities expanded to encompass most aspects of energy production. With each of these events Auerbach's division grew dramatically. Once almost marginal, ecology and environmental science became one of the laboratory's most visible activities.[87]

These events reflected changes in American political priorities. In the mid-1960s the context of federal science shifted as the government began insisting that research be directed toward certain new national goals, including environmental protection. The American contribution to the IBP was one manifestation of this, as it was intended by Congress to provide the personnel and tools for a rational, comprehensive approach to environmental problems. In the 1970s, however, belief in rational management was replaced by a renewed reliance on negotiation, compromise, and brokerage of competing interests more typical of a pluralistic political system. This implied a shift in the role of experts: no longer an alternative to these "political" processes, they became instead participants, contributing knowledge that could be considered objective and therefore useful in decisionmaking and in dispute resolution. In response, research became increasingly tied to the immediate requirements of government and commercial interests. More and more, ecological and environmental research, at Oak Ridge and elsewhere, became linked to specific environmental problems and controversies. Ecosystem ecologists, however, were not immediately able to fulfill these roles. Accordingly, they developed new roles and justification for their research, and scientists with other approaches became more prominent in some areas of environmental affairs.

As a result of these trends, ecological research at Oak Ridge changed in structure. Once their work was perceived as relevant to societal concerns, ecologists became subject to greater outside direction of their work.[88] As a consequence, a single, relatively unified research program was transformed into many narrowly defined research projects and bound to specific environmental issues. Ecosystem ecologists responded to outside direction by protecting their autonomy, particularly by attracting support from the NSF. As they did so, their work grew closer to the concerns of academic ecologists. While ecologists claimed that lasting solutions to environmental concerns and anticipation of future problems depended on ecosystem theory, such theory was not often applicable within the political and economic constraints governing management of actual environmental concerns.

Instead, new areas of research, such as impact assessment, were created to deal with these concerns. Emphasizing quantitative prediction, and responding to the requirements of sponsoring agencies, these specialties were better able than ecosystem ecology to fulfill the demand generated by environmental policies and regulations for "objective" scientific knowledge. Focusing on specific aspects of the environment, such research was also more compatible with the change that had occurred in the role considered appropriate for government, from rational management of the environment and society as a whole to management and resolution of conflict between specific groups within society. Specialists in these fields came to identify themselves as "environmental scientists," a fact Auerbach acknowledged in 1972 by renaming the Ecological Sciences Division, the Environmental Sciences Division. Many of his staff members, with expertise in geology, toxicology, and other disciplines, did not consider themselves ecologists, and wanted a label more consistent with their professional identity.[89]

The permutations of ecological expertise at Oak Ridge during the 1970s—the development of ecosystem ecology by the IBP, its separation from environmental sciences more responsive to administrative and regulatory priorities, and the deferral of much of its authority to these sciences—reflected the impact of changing ideas about society and the environment, and about the appropriate role of government and experts. These changes were relatively recent, with their greatest impact being felt in the 1970s. In the next chapter I consider ecological research that addressed some issues that had been of concern since the progressive conservation era, and the origins of one of the first American science-based natural resource management agencies.

6

Forest Experiment and Practice at Hubbard Brook

In the White Mountains of New Hampshire lies an oval valley logged early in the twentieth century but now covered by a tall, still-growing forest dominated by sugar maple, beech, and yellow birch, and nearly 100 species of shrubs and herbs. Nearly as many species of birds, as well as snowshoe hare, red fox, beaver, black bear, and deer, have been seen in the valley. The landscape is drained by fast-flowing Hubbard Brook, which carries small quantities of soil, organic matter, and stones along a rocky bed as the valley slowly erodes and is swept downstream.

Scientists have been coming since 1963 to study the forests, streams, and lake within the Hubbard Brook watershed. The study was begun by F. Herbert Bormann and Gene E. Likens, two ecologists then at Dartmouth College, with the assistance of Noye M. Johnson, a Dartmouth geologist, and Robert S. Pierce, a U.S. Forest Service hydrologist. By 1989 it had become one of the largest ecological research studies in the United States, with more than one hundred scientists and students gathering that year for the twenty-sixth annual meeting of the Hubbard Brook Ecosystem Study.

Unlike the Nature Conservancy of Great Britain and the Oak Ridge National Laboratory, the Hubbard Brook Ecosystem Study is not located within a single building. It is, rather, a cooperative arrangement between researchers of several universities and the Forest Service, working at a site provided by the Forest Service, and funded primarily by the National Science Foundation. Dominated by university scientists, the Hubbard Brook study also differs from the Nature Conservancy and Oak Ridge in its lack of official responsibility for research on environmental problems (although this is not the case for the Forest Service researchers). It has, nevertheless, contributed to debates over several environmental issues while providing ecologists with a steady flow of information and insights into aquatic and terrestrial ecosystems. In recent years the Hubbard Brook study has become especially widely known for its contribution to the acid rain controversy. But another environmental issue had, throughout its first two decades, an even more significant impact on the study. This was the ecological implication of forest cutting.

ORIGINS

Although the Hubbard Brook Ecosystem Study represented a new research strategy for ecologists, its concept and method had roots in the experience of both Bormann and Likens.

Since the beginning of his career Bormann had studied the ecology of forests and watersheds. As a student of Heinie Oosting at Duke University in North Carolina, where he received his Ph.D. in 1952, he learned to see the temporal dimension of a plant community, viewing through a Clementsian lens the interaction of plants, soil, and climate that results in succession. In his doctoral research he devised experiments to test the significance of various factors in determining the role of two tree species, the loblolly pine and the sweetgum, in the transformation of abandoned fields into forest. It was a promising start to his career; his report won the Ecological Society of America's Mercer Award for the best paper of 1953.[1]

There were close ties between foresters and botanists at Duke. Bormann had received assistance from members of both the department of botany and the School of Forestry. He also visited the nearby Coweeta Hydrologic Laboratory to learn about its work in forestry and hydrology.[2] After graduation he went to Emory University in Georgia. There, he often took his students to Coweeta to show them the watershed experiments, which he considered an excellent method to study the overall characteristics of an entire forest. Given his work with these experiments and his predilection for comprehensive approaches to problems, he was receptive to Eugene Odum's 1953 text on ecosystem ecology, just as Stanley Auerbach had been, although for somewhat different reasons. His own research in plant ecology and physiology also demonstrated an interest in experimental tests of the observations of other ecologists.[3]

In 1956 Bormann moved to Dartmouth College in New Hampshire. With the support of the National Science Foundation and the Atomic Energy Commission he used radionuclides and dye to trace the movement of substances between adjacent root systems.[4] He had not yet formulated a way to study experimentally entire ecosystems with the limited facilities at Dartmouth. But he continued to use an ecosystem perspective in his teaching, relying in part on material he had gathered at Coweeta.

By the end of the decade another way had occurred to him. He had heard of the work by European ecologists on the downslope transport of nutrients by water. William Pearsall, for example, had described how rainwater washes humus and other forms of nutrients out of highland soils.[5] As Bormann recalled, he then realized while hiking beside a mountain stream that this "leaching" or "flushing" (as the British called it) of nutrients could be combined with hydrological information, with which he was familiar from his experience at Coweeta. By measuring stream chemistry and water quantity, the nutrient output of an entire watershed could be determined.[6] Such information could then provide insights into the forest ecosystem.

He already knew of a suitable site for such research: the Hubbard Brook Experimental Forest, a U.S. Forest Service property about 100 kilometers northeast of Dartmouth. From his visits there with his students he was aware that it was equipped to measure streamflow and precipitation. It was also on bedrock believed to be impermeable, preventing deep-seepage, which is impossible to measure. And it was a protected area, safe from interference. Seeking collaborators, he explained the proposed project to his geologist colleagues. They were enthusiastic: measurement of input and output of minerals from the watershed would be of considerable geological interest, as it would permit estimation of rates of erosion, soil formation, and supply of nutrients. He then outlined this concept of a "joint ecology-soils-geochemical problem" to Robert Pierce, director of the Hubbard Brook forest.[7] Pierce agreed to cooperate and to provide the several years of hydrologic data that had been accumulated.

The Forest Service had established the Hubbard Brook Experimental Forest in 1955 to be its major site for research into watershed management in New England. Its immediate purpose was to study the effect of the forest on the quantity of stream water. The Pemigewasset River, into which Hubbard Brook flows, was erratic: often flooding in the spring and almost drying up in the summer, although rainfall was distributed fairly evenly throughout the year. A second objective was to study the relation of water quantity to erosion.[8] Accordingly, the Forest Service installed precipitation collectors, weirs, stream gauges, and flow recorders to monitor the flow of water into and out of six small watersheds within the experimental forest. Pierce and his colleagues also began studying other factors affecting spring floods, including snow accumulation and soil freezing.

Interest in the influence of forests on water supply and flooding had developed even before the Forest Service existed. In 1870 a commission examining waterpower in the White Mountains had concluded that "mountain slopes would, if deprived of vegetation, hasten the discharge of water to such an extent as to render many streams uncontrollable at *one* period of the year and unreliable at another. The wood mosses and fungi which abound in all our dense forests form a spongy mantle to inhale moisture in wet weather and exhale it in dry."[9] Several years earlier George Perkins Marsh, author of the classic work *Man and Nature,* had come to a similar conclusion regarding his home state of Vermont: clearing its forests, he believed, had caused "torrential" floods that swept its streams each spring and autumn, turning them into roily masses of "mud and other impurities."[10]

In 1909 the Forest Service began setting up experimental watersheds to study the effects of land management practices on streamflow. One of the earliest such studies was within fifty kilometers of Hubbard Brook. In 1911 the flows of two tributaries of the East Branch of the Pemigewasset River, one draining a deforested watershed, the other an intact forest, were measured. This comparison confirmed what was already widely believed, that forest cover strongly affected streamflow. In the late 1920s and 1930s concern about erosion provided another impetus for watershed studies. Soon, however, forest watershed studies began to be seen as less essential. By the late 1940s the Soil Conservation Service had assumed leadership in flood control, and watershed experiments, not only of this service but also of the Forest Service, became increasingly devoted to agricultural rather than forest management. Interest in forest watershed studies declined further as the Forest Service began in the 1960s to shift its responsibility away from streamflow regulation and erosion.[11] At this time the service had only a limited program of watershed research at the Hubbard Brook forest.

With Bormann having attracted the interest of both geologists and Pierce of the Forest Service, a small watershed study began in 1961–62, when James Steidtman, a geology student, studied the movement of iodine into and out of a Hubbard Brook watershed.[12] But with this exception, the geologists chose not to participate. However, Gene Likens, who arrived at Dartmouth in the fall of 1962 from the limnology program of the University of Wisconsin, was very interested in a watershed study (see fig. 6).

Since the work of Edward Birge and Chancey Juday in the 1890s, Wisconsin limnologists had sought to understand lakes through inductive generalization from data accumulated in hundreds of surveys. But in the late 1940s, when Arthur Hasler became director of the program, they began a new approach, seeking, in Hasler's words, to "apply the methods of the experimental laboratory to the field."[13] In their first of several successful experimental manipulations of lake ecosystems, Hasler and his colleagues closed off a narrow channel that linked Lakes Peter and Paul. Lime was then added to Lake Peter to raise the pH, and the resulting chemical and biological changes were examined and compared with conditions in Lake Paul.[14] As one of Hasler's students, Likens had himself conducted an experiment in which he followed the movement of radionuclides within lakes. This limnological study reflected his interest in studying the behavior of entire ecosystems. His reading of the second edition of Odum's textbook confirmed it, because it showed him that through ecosystem study he could combine his interests in plants, animals, and their environment. Likens met Bormann when a short-term teaching ap-

Fig. 6. F. Herbert Bormann, Gene Likens, and Robert Pierce (courtesy of Gene Likens)

pointment brought Likens to Dartmouth just before he graduated. They discovered their common interest in a biogeochemical approach that would focus on the movement of matter within ecosystems. Likens returned as a faculty member in 1963. He then met Noye Johnson, a geochemist who had just joined the Dartmouth geology department, and solicited his interest in a watershed study.[15]

Between 1962 and 1966 Bormann and Likens, with Johnson, developed and refined their plan to use small watersheds to study ecosystems. They submitted their initial proposal to the National Institutes of Health of the Public Health Service (PHS). Because they proposed to measure stream water chemistry, their choice of agency was perhaps logical, as the PHS was at this time responsible for monitoring water pollution. The scientists suggested that their method of watershed study could eventually be used to evaluate the behavior of pollutants released into watersheds. The proposal was, however, rejected. They then submitted a similar proposal to the NSF, which approved it, providing support for three years.[16]

The project began slowly, as both Likens and Bormann were completing other projects.[17] Bormann also spent part of 1963 and 1964 at the Brookhaven National Laboratory as a visiting scientist. In 1963 they began taking samples for chemical analysis of precipitation and streamflow from several watersheds within the Hubbard Brook Experimental Forest. With these samples began a record of precipitation and stream chemistry that extends to the present day. In January 1967 they published a paper in *Science* explaining to the scientific community what they could do with this data. They sought not only to convince others of the merits of watershed ecosystem study but to motivate others to follow their example: "Studies similar to these at HB could be established elsewhere in the United States. There are thousands of gaged watersheds operated by private and public interests, and some of these must meet the proposed requirements. On selected watersheds, cooperative studies could be made by the agencies or organizations controlling the watershed and university-based investigators interested in biogeochemical cycling. Just such cooperation, between federal agencies and universities, has been urged by the Task Group on Coordinated Water Resources Research."[18]

Initially, their conception of the terrestrial ecosystem was quite simple: it was composed of a biotic component (living organisms) and an abiotic reservoir. Substances continually move from the abiotic reservoir to the biotic component, circulate, and eventually return to the abiotic reservoir.[19] By 1967 the researchers had refined this view. Nutrients, they now explained, were within four compartments of the watershed ecosystem: in the atmosphere; in organic material, both living and dead; in the soil, available for eventual uptake by plants; and locked away within soil and rock minerals. A variety of biological and physical mechanisms moved nutrients between these compartments.[20]

It was most important, they argued, to understand that ecosystems are open systems linked by the movement of materials to other ecosystems. They believed, and other ecologists agreed, that study of these linkages was essential to a complete understanding of ecosystems. As Ovington had recently noted in England, "Long-term changes in woodland soils cannot be interpreted solely in terms of nutrient incorporation within the organic mass and loss from the soil, but account must be taken of the complete ecosystem dynamics, particularly the external flow of nutrients into and out of the woodland ecosystem."[21]

There were, Bormann and Likens explained, three types of inputs: geologic (entering the ecosystem at or below ground level, through erosion and colluvial movement); meteorologic (entering the atmosphere as dissolved substances in precipitation, as dust or other particulate matter, or as gases entering through biological processes such as nitrogen fixation); or biologic (material transported into the ecosystem by animals). Outputs could be similarly classified into these categories. These inputs and outputs, while necessary to understanding ecosystems, had, they noted, been neglected by ecologists. In particular, little study had been made of the role of erosion and weathering.[22] Similarly, while scientists had studied the chemistry of rainfall and stream water within drainage basins, as well as the hydrology of small watersheds, there had been little integrated study of the ecology, hydrology, and biogeochemistry of a single ecosystem. Their study, therefore, would not be guided by questions deduced from existing theory but through induction—by determining, for example, the effects of ecosystem processes on streamflow, or the significance of precipitation chemistry to these processes.[23] In effect, their study would colonize an "intellectual 'no man's land' between traditional concepts of ecology, geology, and pedology."[24]

The ideal vehicle for this colonization, they argued, would be a small watershed with limited paths for the input or output of water or matter. If underlain by impermeable bedrock, as Hubbard Brook was believed to be, there was no possibility of deep seepage out of the watershed. Watersheds are by definition separated by heights of land, and so landslides and other forms of downslope transport of rock and soil could not transfer materials in or out. And if the watershed was within a larger vegetation unit, like the forests around Hubbard Brook, then movement of materials into and out of the watershed by animals could be disregarded.

In a watershed where these assumptions were valid, water could enter only as rainfall and leave only as runoff, by evaporation, or by transpiration. With precipitation and streamflow gauges the flow of liquid water into and out of the watershed could be measured. Through chemical analysis of the water entering and leaving the watershed nutrient budgets could be determined, indicating the input and output of each nutrient and the net gain or loss of each by the ecosystem. "Nutrient input and output are directly related to the amounts of water that move into and out of an ecosystem," they noted later, "while temporal and absolute limits of biogeochemical activities within the system are markedly influenced by the hydrologic regime."[25] These budgets could then be used as a baseline for studies of forest and stream ecology, or of the effects of human activities on the ecosystem. And a water budget could be calculated, as forestry researchers had already been doing, to provide an estimate of evaporation and transpiration.

The focus on nutrient movements, shared by ecologists at Hubbard Brook and Oak Ridge (and by Ovington), reflected rapidly growing interest in ecosystem ecology at various research sites. However, while the Oak Ridge cesium forest experiment focused on nutrient movement within the ecosystem, at Hubbard Brook there was interest in movement between ecosystems. These different approaches had contrasting methods: radiotracers (at Oak Ridge) and watershed monitoring (at Hubbard Brook), which were suitable for measuring different aspects of ecosystem dynamics.

If it was also assumed that the forest was at the climax stage of succession, then geologists could use nutrient budgets. With the ecosystem no longer accumulating biomass, net losses of any nutrients would have to be made up through weathering. Johnson applied the budget data as it became available to measure this otherwise difficult-to-quantify geological process.[26]

The design of the Hubbard Brook study combined the expertise and interests of its participants. It also reflected Bormann and Likens's ambition to explore an approach to ecology in which study of the flows of chemical elements and compounds within and between ecosystems would herald a rigorous, quantitative ecological science based on the principles of chemistry and physics. Likens recalled that both he and Bormann hoped to use their quantitative approach to replace the "butterfly collector" image that they felt ecologists had within much of the scientific community.[27] As we have seen, this ambition also motivated ecologists in the Nature Conservancy and at Oak Ridge. At all three sites (especially at Oak Ridge) this goal encouraged them to adopt an ecosystem approach as a means through which a more rigorous form of ecology could be brought into being.

Their colonization of an "intellectual no-man's land"—achieved through linking the study to forestry, soil, geological, and ecological issues—was an effort to establish the ecosystem not only as a distinctive unit of study for ecologists but as a common basis for several disciplines concerned with different aspects of the environment. This effort was much more evident at Hubbard Brook than in Nature Conservancy ecological research at Merlewood, or at Oak Ridge. Bormann and Likens's ambition to combine disciplines likely reflected their own backgrounds and interests, as well as their need to attract potential collaborators. With Dartmouth having only a small ecology program, those collaborators had to come from other disciplines.

THE FIRST WATERSHED EXPERIMENT

In their 1963 proposal Bormann, Likens, and Johnson had anticipated that once they had a basic understanding of the watersheds' biogeochemistry they could study how it would be affected by such manipulations as clearcutting, burning, or application of insecticides. By 1963 they were prepared to supply only a rough sketch of the proposed experiments, which they noted were to be performed by the Forestry Service.

When the study was two years old, however, Bormann and Likens themselves began an experiment. They selected a 15.6-hectare watershed known as Watershed Two and arranged for all its trees to be cut down during November and December 1965. The trees were left where they fell, and the soil was not disturbed. Herbicide was then applied to prevent regrowth of vegetation. This treatment continued for three years, after which the plants were permitted to reestablish themselves.

Bormann and Likens suggested that vegetation, by absorbing nutrients from the soil for its own growth and metabolism, helped retain nutrients within the ecosystem, ensuring continued productivity. By cutting the trees and measuring subsequent nutrient losses from the ecosystem, the role of trees in regulating nutrient cycling and export could be measured. Herbicide treatment would ensure that measurement of nutrient loss would not be confounded by new plant growth and uptake of nutrients from the soil. The use of herbicide and the practice of leaving the logs in place indicated that the experiment was not designed to duplicate conventional commercial logging. The primary aim was to understand the functioning of an intact forest, not the effects of cutting one down.

Bormann and Likens were eager to begin a watershed experiment. As I have noted, Bormann had demonstrated his interest in experimentation. Experiments, he had argued, could provide more reliable—and often more easily obtained—results than could descriptive studies.[28] Likens, who had recently completed his own ecosystem experiments in Wisconsin, was also convinced of their value. Their attitudes toward experimentation recall the

desire of British ecologists to obtain access to nature reserves, in part to initi-
ate field experiments. In both cases experimentation was seen as providing,
through control and manipulation of selected environmental variables, information concerning ecological processes that could not be obtained through
descriptive studies. It is also interesting, however, to compare these attitudes
with Auerbach's initial reluctance to begin field experiments at Oak Ridge.
His concern that field conditions would generate results perceived by physical scientists as less rigorous hints at the potential role of other disciplines in
evaluating ecological experiments in which the control of variables, or replication, would be more difficult. In contrast, this was of less concern to Hubbard
Brook ecologists (and Nature Conservancy ecologists), as they could expect
their results to be evaluated by other ecologists, not by scientists from other
disciplines.

Bormann and Likens's interest in experimentation showed that they understood funding realities. An experimental study was more likely to receive
support from the National Science Foundation than was one based only on
monitoring. Bormann, who had been preparing successful proposals to the
NSF since the early 1950s, knew this. So did Likens, having studied with Arthur Hasler. Hasler, notably successful in arranging funding for Wisconsin
limnology, justified his departure from his predecessors' descriptive surveys
by noting that experiments were more likely to receive support. "I could see
that it was presumptuous to try to make a reputation by following in the Birge
and Juday research tradition," he recalled in 1985. "Moreover, you couldn't
get money for that type of research—recording the environment."[29] Likens
had clearly absorbed this lesson, and he pushed hardest to begin the experiment. Perhaps reflecting the Forest Service's declining interest in watershed
studies, Pierce and his colleagues were more reluctant. Their concern was
with forest hydrology, and it was not apparent that the experiment would produce significant insights in that area, nor that enough baseline data had been
collected.[30] After initial hesitation, however, they agreed to cooperate. Their
support would prove essential to the study.

As Bormann, Likens, and their colleagues continued to measure the quantity and chemical composition of stream water and precipitation, they observed immediate changes in the export of water and elements from the
watershed after the forest was cut. Streamflow increased 39 percent in the
first year and 28 percent the second year, above that expected if the watershed
had remained undisturbed. Most of this increase occurred in the summer,
with more than five times the normal streamflow the first summer.[31] These results were not unexpected: watershed clearings at Coweeta and elsewhere
had resulted in increased streamflow. It had become a basic principle of

watershed management that removal of trees would result in a greater flow of water. Pierce and his forestry colleagues interpreted the extra summer flow as water that would have been lost through transpiration during this season of rapid plant growth.[32]

The cutting also affected the loss of particulate matter by the watershed. They found, after a delay, a fourfold increase in matter suspended within the stream or swept along its bed. This, they argued, demonstrated the importance of vegetation in reducing erosion. It did this by decreasing the amount of runoff, particularly at times of flooding, through small dams (of leaves, twigs, and other organic matter that slow the flow and allow particulate matter to accumulate), and through roots and leaves that bind and protect stream banks from rushing water. With the forest cut, these obstructions to erosion were worn away.[33]

Erosion, like increased streamflow, had been anticipated by forestry researchers. Bormann and his colleagues were well aware of the insights of E. A. Coleman and other foresters on the significance of forest biomass in reducing erosion.[34] In interpreting their results in light of previous research on watershed hydrology and sediment yield they aimed to demonstrate that forestry research could make an important contribution to the Hubbard Brook study and to ecological research generally. As they later noted, ecologists could learn much from foresters: "It is interesting to note that this most basic and important perturbation to ecosystem stability, erosion, long known to agriculturalists and foresters, hardly enters the current ecological debate on stability of ecosystems."[35]

Awareness of the ideas of foresters probably came less from Forest Service researchers at Hubbard Brook than from their own (particularly Bormann's) experience and education. In the 1960s Pierce and his colleagues, James Hornbeck and Anthony Federer, did not work closely with Bormann and Likens. While Pierce's contributions (including permission to use Forest Service land and hydrologic data, and coordination of research and watershed management activities within the experimental forest) were crucial, Bormann and Likens themselves planned the research and prepared the NSF proposals. They did not begin until 1969 to jointly plan some research, when Pierce prepared part of the proposal for research from 1969 to 1971.[36]

Bormann and Likens saw their most dramatic results not in increased streamflow and erosion but in data not previously collected by forestry researchers: stream water chemistry. In spring 1966, five months after the cutting, the export of dissolved solids rose dramatically. Over the next two years the amount exported was six to eight times as much as from an undisturbed watershed. The concentrations of most ions greatly increased, including calcium, magnesium, potassium, and sodium. Nitrate (NO_3-), however, led all of these: the first year its concentration was 41 times higher than that of an undisturbed watershed, and the second year, 56 times higher.[37] Clearly, the watershed ecosystem had lost much of its ability to retain nutrients.

These results could be explained only partly in terms of the existing understanding of watershed ecosystems. Several years of monitoring the flow of water and elements from undisturbed watersheds had confirmed that the forest was well able to minimize its loss of dissolved and particulate substances. These data, as summarized in an empirical model developed by Noye Johnson, showed that the concentration of dissolved substances in stream water, and hence their loss from the watershed, was a function of stream flow. Johnson's model could therefore predict loss of nutrients from an undisturbed watershed if the stream flow was known. This meant that forests regulated the loss of dissolved nutrients by regulating the flow of water out of the ecosystem.[38]

It had been expected that this model could also predict nutrient export from the experimental watershed once allowance was made for the loss of demand for nutrients by the vegetation, and for other biological factors.[39] But the researchers quickly found that the model could not predict the loss of nitrate and other ions from the watershed. Another process had to be responsible.

Bormann and Likens, assisted by William Smith, a soil scientist at Yale, quickly identified this process as nitrification. In an undisturbed forest, nitrogen incorporated within vegetation converts to ammonium when the vegetation dies and decomposes. This ammonium may then be taken up by new plants or may be retained within soil or organic matter on the forest floor. If the vegetation is eliminated, however, nitrification is greatly accelerated, as chemoautotrophic or heterotrophic microbiota on the forest floor rapidly oxidize the ammonium, producing great quantities of nitrates. Highly soluble in water, the nitrates can then be flushed rapidly from the forest floor while creating conditions for increased mobilization of various ions.[40]

This discovery of nitrification called for a major research effort. "The turning-on of the process of nitrification after clear-cutting our forest," Bormann recalled, "was totally unexpected—a truly major event in the biogeochemistry of the ecosystem. This opened a wholly new aspect of research, and other things had to give way."[41] Likens has described the discovery of nitrification as the most significant result of the Hubbard Brook study.[42] As the researchers quickly realized, nitrification, occurring within a soil environment previously considered too acidic for this process and serving as the linchpin of nutrient loss from a disturbed ecosystem, was the dramatic, unexpected discovery that could demonstrate the utility of watershed experiments.

Bormann, Likens, and their colleagues drew much attention to nitrification, featuring it in their analysis of the watershed experiment and reporting the discovery itself in the journal *Science*. New research projects were also initiated. For example, Tony Dominski, a Yale doctoral student, compared the chemical transformations of nitrogen in Watershed Two with undisturbed watersheds to determine how changes in the soil resulting from forest cutting initiated or accelerated nitrification. Other studies focused on aspects of the regulation of nutrients and water movement within the forest. For example, James Gosz studied the role of litter decomposition on the forest floor in the cycling and retention of nutrients within the ecosystem.[43]

The Forest Service, and particularly Pierce, contributed to the success of the watershed experiment. The service, as I have noted, had given permission for the experiment to be conducted in the first place and had provided the necessary hydrologic data. These contributions were apparently reciprocated to the satisfaction of the Forest Service; officials were said to be pleased with an early paper explaining the rationale and initial results of the study.[44] But a problem appeared in 1968 with the first published report of the increased loss of dissolved nutrients that occurred after the cutting. The paper appeared in *Science* under the title "Nutrient Loss Accelerated by Clear-Cutting of a Forest Ecosystem." Forest Service officials of the Northeastern Forest Experiment Station in Upper Darby, Pennsylvania, were uneasy about this title. As Pierce explained, the problem was the term *clearcutting*, more typically used to refer to conventional logging practice. Although the article described the experimental treatment, including the use of herbicide, Forest Service officials were concerned that others might assume that the study was reporting the effects of typical forest harvesting practices.[45]

As a result, the researchers afterward were careful to distinguish the experiment from forestry practice. Discussing the preparation of another paper, tentatively titled "Effects of Forest-Cutting on Hydrologic and Nutrient Budgets in the Hubbard Brook Watershed-Ecosystem," Pierce suggested to Likens that *cutting* be replaced by *denudation*.[46] Although Likens and his colleagues chose not to take this advice, the title eventually used, "Effects of Forest Cutting and Herbicide Treatment on Nutrient Budgets in the Hubbard Brook Watershed-Ecosystem," did, by mentioning the herbicide, highlight the feature that most distinguished the experiment from forestry practice.[47]

The Forest Service decided to begin its own studies to determine whether nutrient losses would also occur as a result of conventional forestry practice.

Pierce, Federer, Hornbeck, and Wayne Martin designed and conducted these studies, with Bormann and Likens assisting. In 1970 another area within the experimental forest was clearcut by commercial loggers, who used their normal techniques, including skid trails and removal of timber. This study was begun quickly, before a weir and stream gauge had been installed, and so streamflow quantity could not be measured. Water samples were taken, however, in an attempt to measure nutrient export. In a second study, the researchers measured nutrient export from several commercially clearcut sites throughout New Hampshire and Maine. All these studies found substantial loss of nitrates from the watersheds, though generally less than in the original deforestation experiment.[48]

The largest study was on Watershed Four, located near the watershed used in the first experiment. Pierce's aim in this, as in the other studies, was to measure the amount of nutrients exported as a result of usual logging practices. At first he and his colleagues planned to clearcut the entire watershed. But they then decided that clearcutting would likely soon become obsolete, as would a study of its effects. Instead, they chose to study "stripcutting," a technique that many foresters expected to become more important than clearcutting. Stripcutting was considered to have several advantages, including faster natural regeneration.[49] They began by dividing the watershed into horizontal strips, each twenty-five meters wide. One-third of the strips were cut in 1970, 1972, and 1974. In addition, a strip of trees was left uncut on each side of the stream, also corresponding to standard forestry practice. They measured streamflow, nutrient losses, and natural regeneration in small plots throughout the watershed.

The differences between Watershed Two (clearcut, treated with herbicide) and Watershed Four (stripcut, with vegetation buffer along the stream) mirrored the different objectives of the ecologists and the forestry researchers. Bormann and Likens were not studying the environmental impacts of a specific forest cutting technique. Their experiment had been designed to modify one aspect of the ecosystem—the vegetation—in order to better understand the functioning of intact forest ecosystems. Their primary interest was not the loss of nutrients from the watershed but what this loss indicated about the importance of vegetation in maintaining ecosystem stability.[50] Measurement of nutrient loss, as a symptom of the overall condition of the ecosystem, could also help provide the knowledge base for efforts devoted to balancing preservation of the environment and the multiple demands for resources within that environment. As they argued in 1969, "Given the basic abiotic and biotic complexity of land, the phenomena of succession and retrogression, a multiplicity of managerial goals, and a desire for more efficient use of the land, it is obvious that some theoretical framework upon which we can assemble and interrelate the diverse compartments is a necessity. . . . The ecosystem concept provides this framework."[51] The theoretical framework they referred to was central to their own study: the flow of energy and materials. In effect, these flows could provide a common language for study of the interactions between those factors shaping the landscape—the processes of nature and the demands placed on them by humans—not otherwise directly comparable. This represented a further development of Bormann and Likens's view of the ecosystem as a common basis for several environmental disciplines, not only ecology.

Pierce and his colleagues, in contrast, had much more specific goals. They wanted to duplicate a particular type of forestry practice and determine how it affected a specific aspect of the environment—nutrient loss—because this had recently become of wide concern.

But this distinction—between forestry research applied to a specific problem and ecological research aimed at a general understanding of ecosystems—was lost on many. One reason was that popular accounts of the first experiment misrepresented it. An item in *Science News,* for example, described it as an "effort to predict the effects of forestry practices on the ecology," and it noted, incorrectly, that all the trees had been removed.[52] In part as a result of such misunderstanding, Hubbard Brook became enmeshed in a national controversy over the environmental effects of logging, particularly the impact of clearcutting. In several instances environmentalists used data from the Watershed Two experiment to support their contention that clearcutting of an area depleted its nutrients.

A particularly visible stir was created on April 6, 1971, when Robert Curry, professor of environmental geology at the University of Montana, testified to a Senate Committee examining clearcutting practices. He told the senators of "forest soils as nutrient reservoirs which take tens of thousands of years to form and which are now being lost through faulty logging practices at rates hundreds to thousands of times faster than their formation." Chief causes were clearcutting and some forms of selective cutting, which set up a chain of events leading to loss of nutrients, even the "sterilization" of the soil. Clearcutting, he said, was "killing the soil."[53]

Curry aimed to shock, and he succeeded. One senator told him that "yours is indeed the most challenging testimony that we have received." And the culprit was also clear; as Senator Frank Church said, "If you are right, Dr. Curry, the indictment against the Forest Service is a very severe one." But most important to this account is that at the core of Curry's evidence were the just-published results of the Hubbard Brook experiment. Its demonstration of accelerated loss of nutrients could, he believed, be applied to conditions across much of the nation. In a week marked by a Senate bill proposing a two-year moratorium on clearcutting, intense coverage in the newspapers, and even a Walter Cronkite television special on clearcutting, his testimony was guaranteed wide attention.

The reaction of Forest Service officials to this public condemnation of their practices, fueled by results from a study done on their own land, can be imagined. At the very least it apparently confirmed their fears of two years before, that referring to the experiment as a "clearcut" would lead to misinterpretation. As another researcher told Pierce sympathetically, "I share the regrets that you and your co-authors have concerning the mis-use and wild extrapolation of the results. . . . You and your co-workers have taken some heat about the 'artificial' or 'extreme' treatment of [Watershed Two results] and you may well be a little sensitive about this point and the continued emphasis upon it."[54] Evidently forestry researchers, while aware that the Watershed Two experiment did not duplicate normal forestry practices, were critical of its "artificial" and "extreme" design, which they believed made clearcutting appear more damaging than it really was.

This controversy was just as unwelcome to Bormann and Likens, and they disassociated their experiment from the conclusions others were drawing regarding the environmental impact of forestry. Both declined to testify before Congress. "This experiment was not designed to simulate a practical forest operation," they stressed to a gathering of foresters, "but was solely for research purposes."[55] This decision to stand back from the controversy reflected their perception of their role as ecologists: to provide ecosystem theory that could be the basis for environmentally sound forest management. Although, as they themselves noted, their research could reveal aspects of ecosystems that should be considered by applied scientists and managers, such as nutrient cycling, they would not become directly involved in specific controversies. "We figured we should be correct with what we do and what we say about our place, and if others were going to use it, that's up to them," recalled Bormann, "But it's not our job to be kind of policemen of the scientific literature."[56] With many ecologists concerned during this Age of Ecology about the reputation of their field as a fully scientific, objective discipline, their reticence may been an attempt to safeguard the credibility of the Hubbard Brook study.

Thus, Bormann and Likens aimed to contribute to forest management practices while not participating in controversy. Instead, their contribution would be made in less confrontational arenas, through provision of a theoretical understanding of forested ecosystems. A comparison can be made with contemporaneous events at Oak Ridge. At Hubbard Brook, ecologists could choose not to be directly involved in controversy; in contrast, Auerbach, because of the AEC's regulatory responsibilities, was obliged to involve his Ecological Sciences Division in controversies concerning the environmental impact of nuclear power plants and other facilities on the Hudson River and elsewhere.

As the clearcutting controversy simmered in Washington and elsewhere, Bormann and Likens, with assistance from Pierce, prepared a new proposal for National Science Foundation funding for 1971 to 1973. As usual, the NSF decided to continue its support. But accompanying this decision were a number of conditions and qualifications. The most significant concerned several projects relevant to forest management, including the stripcutting experiment, begun the previous year, which Bormann and Likens proposed continuing with NSF support. Charles Cooper, program director for the NSF's Ecosystem Analysis Program, explained why the NSF disapproved of funding such a project:

> We . . . do not consider that this research falls within the responsibility of NSF. The reaction might be somewhat different if it were shown that the results would have general applicability to a wide range of forest conditions outside Hubbard Brook (although even here there is a real question about the appropriateness of NSF involvement). The NSF role would be even more obvious if there were provisions to test a general model of wide applicability (or even of local application) derived from understanding of ecosystem properties. This I recognize is still beyond the capability of ecological theory. The conclusion must remain, however, that we cannot authorize the use of NSF funds for the research associated with commercial strip cutting.[57]

The NSF's priority, Cooper was explaining, had to be research contributing to development of general ecological theory. Study of the effect of a specific forest management technique within a single watershed did not promise to do so. Cooper also requested more information on what the Forest Service was contributing to the study, and "just what NSF is expected to pay for." This suggests that the foundation feared having to support research that was the responsibility of the Forest Service.

Bormann and Likens strongly opposed these conditions. "We seriously question the policy of the National Science Foundation as stated in your letter," they responded. "The Foundation questions the appropriateness, for NSF support, of the very heart of our study. Not only does this part of our study contribute to basic understanding of ecosystem processes but it contributes most to the solution of very pressing national environmental problems. We believe this is a short-sighted policy on the part of the Foundation and not in the best national interests." They could not, they insisted, break the project down into Forest Service and foundation components; to do so would deny the mutually beneficial relationship enjoyed by foresters and ecologists: "It should suffice to say that the Forest Service is responsible for a major part of the Hubbard Brook Study and that they have made a substantial investment in the overall project. As a consequence of their investment, we think that the money spent by NSF at Hubbard Brook has a higher scientific return per dollar than any comparable project in the country."[58] They also appealed to the director of the NSF, William McElroy, who affirmed that the foundation could not attach conditions to its grants or forbid certain research as unsuitable for NSF support if it had been successful in the peer review process.[59] This protest worked, and the conditions imposed by the NSF were lifted.

This controversy was a minor skirmish; cooperation between Forest Service researchers and ecologists continued, and even intensified in the case of some research topics of mutual interest. What was significant, however, was that it highlighted the obstacle to cooperation implicit in the varying objectives of the sponsoring agencies: the management-oriented Forest Service and the basic research-oriented National Science Foundation. These divergent priorities resulted in pressure to separate research devoted to forestry issues from that devoted to basic ecosystem ecology. This pressure was felt in both the planning of research (as imposed by the NSF) and in the presentation of its results, when it became necessary to stress that the first ecosystem experiment was not directly relevant to actual forestry practice.

During the late 1960s and 1970s Forest Service scientists also began several projects in the Experimental Forest on such topics as the hydrology of forested watersheds, particularly the interaction between evapotranspiration and streamflow, the effect of forest management practices on watershed nutrient loss, and the role of snow in the forest ecosystem.[60] Pierce's role as senior Forest Service scientist was to coordinate and manage the service's program, and he contributed to most of its projects at Hubbard Brook.

Increasingly during the 1970s, Forest Service researchers and ecologists cooperated on research projects. They also shared in deciding research priorities through a Scientific Advisory Committee composed of senior researchers and chaired by a Forest Service scientist. In some instances they collaborated on projects. Hornbeck, for example, worked with Likens in studies of snow and acid precipitation.[61] In other projects, the cooperation was more indirect, as, for example, in Wayne Martin's use of chemistry data and methods developed by Likens and Bormann.

For Pierce and his colleagues, cooperation with ecologists was essential to their effort to transfer the ideas and methods of ecology into their research.

The forestry profession came under pressure in the 1970s to replace its emphasis on maintaining a sustained yield of wood with a focus on the more broadly defined environmental values of forests. These values included maintaining the capability of forests to sustain multiple uses, including recreation, nature study, and the protection of wildlife habitat, as well as such new concerns as nutrient depletion and the effects of air pollution on forests.[62] Pierce and his colleagues led the Forest Service in reshaping forestry research to respond to these concerns.

HUBBARD BROOK AND ECOSYSTEM THEORY

In 1966 Bormann moved to the School of Forestry and Environmental Studies of Yale University. Three years later Likens also left Dartmouth, going to the Section of Ecology and Systematics of Cornell University. These moves did not significantly affect their collaboration, however. Although they no longer had adjoining offices, during the summer they were both at Hubbard Brook, and during the winter they met regularly to plan research and cowrite papers.

Yale and Cornell were hospitable to the combination of research and development of practical implications that Bormann and Likens wanted to develop. A second and perhaps more important advantage was that Yale and Cornell, major centers for graduate education, made it possible to involve substantial numbers of graduate students in the study. Dartmouth, in contrast, had few graduate students.[63] Much of the growth of the Hubbard Brook Ecosystem study during the 1970s was the result of the participation of students. Between 1970 and 1980 more than forty students completed doctoral or master's degree dissertations based on research at Hubbard Brook. By involving students Bormann and Likens greatly increased how efficiently and flexibly they could use their funds. Students were often supported by fellowships and teaching assistantships, so only the direct costs of their research needed to be covered by the Hubbard Brook grant. Their research was inexpensive enough that study of some topics not included within the grant could still be supported by it. For example, research on Mirror Lake (the only lake near the experimental forest) seldom has been provided for in the NSF grant, but it has gone on almost continuously since 1967.[64]

In 1968 the deforested watershed was for the last time treated with herbicide. Colonizing plants appeared the following year, and by 1972 vegetation had reached several meters in height. Close study of the growth of this vegetation produced a record "of the pattern of recovery . . . [that is] the most complete account of this type anywhere."[65] Other studies of ecosystem recovery from disturbance accompanied this monitoring. Central to this process, as ecologists studying succession had long known, were the first colonizers of an area, which through rapid growth create an environment suitable to those species that will eventually replace them. Bormann had noted the presence of one of these colonizers, the pin cherry, in many clearings around Hubbard Brook. In 1968 he assigned a doctoral student, Peter Marks, to study how it establishes itself on disturbed land and to identify its role in ecosystem recovery.

Marks found that the pin cherry's rapid growth and uptake of water and nutrients reduced their loss from the deforested watershed, helping to move the ecosystem back toward a stable pattern of nutrient circulation. The full canopy also moderated soil temperature, thereby reducing the supply of soluble ions available for loss in drainage water. Because of these contributions, the pin cherry, like other successional species, had to be considered an integral part of the forest ecosystem. There was, Marks and Bormann stressed, a clear practical implication: these species, while of no economic value, were not weeds. They were a vital part of the recovery of the forest from cutting, and forestry practices that eliminated them would be ecologically unsound.[66]

In reporting the results of the Watershed Two experiment, Hubbard Brook ecologists had, as I have noted, highlighted the potential contribution to ecology of forestry researchers' experience in watershed study. In doing so they were also using this experience to provide additional credibility for their own results. Forestry research provided them with independent corroboration. One demonstration of this came in 1972, when they were preparing their full report on the cutting of the watershed and its subsequent recovery. In May they submitted their manuscript to the editor of *Ecological Monographs*. When it was returned in December (after a delay caused by a "dilatory reviewer") they learned that the manuscript had been accepted but that both reviewers had reservations about its lack of corroborating evidence. As one reviewer complained, "Perhaps the greatest downfall of this paper is the attempt by the authors to make some suggestions on how all ecosystems behave through their detailed observations on a small drainage area. If they insist on expanding their conclusions to cover other than what their data would indicate, then they should have the responsibility to include the extensive literature available on this subject."[67]

The effect of this advice was seen in the final paper, in which they noted several watershed studies by foresters and the consistency of the results of these studies with their own.[68] Such independent confirmation, while important for all scientific experiments, was especially vital for ecosystem experimentation, because their expense prevented repeated testing of results. Their corroboration over a range of conditions by other studies was therefore crucial to their acceptance by the scientific community. Harvesting operations occurring throughout the forests of New England, together with the results of decades of study by forestry researchers, could provide such corroboration relatively inexpensively.

While considerable attention focused on watershed experiments, there was also study of the undisturbed forest, including the distribution and growth of tree species and the undergrowth in relation to environmental conditions and the historical impact of logging, fire, and other factors.[69] Data from these surveys then provided raw material for two studies of the past and potential growth of the forest, one led by Robert Whittaker, the other by Daniel Botkin.

Bormann had long been familiar with Whittaker's work, and he knew of the Brookhaven system of forest dimension analysis that Whittaker and George Woodwell had recently developed.[70] On the basis of measurements of selected trees, this system could deduce trends in forest productivity. When Whittaker applied it to Bormann's data, he found that while the Hubbard Brook forest had a species composition similar to that of a mature climax forest, the accumulation of biomass characteristic of succession was still under way as the valley continued to recover from being logged about fifty-five years before. The assumption made earlier—that the forest was at a steady state, with no net accumulation of biomass—was therefore incorrect.[71]

They also concluded that there had been a substantial (18 percent) decline in forest productivity between 1961 and 1965. These years were marked by drought, which may have caused the decline. But, as Whittaker and his colleagues noted, growth had never declined by so much during the preceding two centuries. Perhaps, as Swedish scientists had proposed, the more novel phenomenon of acid rain could be a factor in reducing forest growth.[72] This suggestion, presented as an apparent afterthought, may in fact have been, for Bormann at least, the study's most significant result. "My objective," he told Whittaker in explaining changes he was suggesting for a joint paper, was to "allow for the equivocations that our new data introduced . . . and yet allow us to go on record that there is some evidence that declining tree growth may be occurring as a result of increases in acid rain. Further, I hope this paper will provide a solid base for an intensive tree ring study in our next proposal. The major objective of that study will be to nail down the declining trend and relate it to particular species and sites."[73]

As this effort continued, a second group was using a very different method to explore the future growth of the Hubbard Brook forest. This began in 1968 with the development of a forest growth model by Larry Forcier and Bormann. This model was predicated on size classes within tree populations and movement of individual trees into large size classes through growth, or loss from a size class through death.[74] The following year Tom Siccama developed an empirical computer simulation based on this model, using forest data collected by Bormann and his colleagues. At this time, with the IBP biome programs getting under way at Oak Ridge and elsewhere, many ecologists were exploring the possibilities of computer modeling. Bormann and Likens were no exception, and they agreed that some modeling should be brought into the Hubbard Brook study. Bormann knew a colleague at Yale who could do this: Daniel Botkin, who had a strong grounding in mathematics and physics and who had already done some modeling of photosynthesis at Brookhaven. Likens knew that IBM was also interested in applying computers to ecology. Bormann and Likens accordingly brought together Botkin, Siccama, and two scientists from IBM—James Wallis and James Janak.[75]

Botkin, Wallis, and Janak used the Hubbard Brook forest data to develop a computer model named JABOWA (JAnak, BOtkin, WAllis). It simulated the growth of individual trees in a mixed species forest, taking into account competition among trees, growth and survival characteristics of each species, and the response of each species to such environmental variables as soil moisture, light, and nutrients. As a model based on individuals, it was therefore very different from models then being developed at Oak Ridge, which focused on the movement of nutrients and energy, and deemphasized the significance of individuals. By 1971 Botkin and his colleagues were able, by simulating these competitive and environmental interactions year by year, to make long-term predictions of forest species composition and tree size.[76] Whittaker's dimension analysis was then applied to the model results to predict long-term trends in forest biomass. The most significant prediction was that as succession continued, the largest trees would begin to die, to be replaced by fast-growing colonizers. Eventually the forest would consist of trees of all ages, and total forest biomass would have declined from its maximum. This suggested that the classical concept of the climax forest, as a uniform and stable forest at maximum biomass, was not valid.[77]

In documenting their model, Botkin and his colleagues did not discuss its application to environmental problems. Botkin, however, went on to apply the model to different combinations of tree species and environmental conditions. In 1977, for example, he simulated the effect of an increase in atmospheric CO_2 on forest growth.[78] By this time Botkin had moved to Woods Hole, Massachusetts, and he no longer identified his model as a part of the Hubbard Brook study.

Both Whittaker's and Botkin's contributions were the product of invitations from Bormann and Likens to apply their expertise to problems encountered in the Hubbard Brook study. They often used this strategy to enlarge their pool of expertise; Bormann listed in 1986 sixteen scientists whom they had recruited in this way. This also stretched the budget, because most of the scientists were not paid to work at Hubbard Brook.[79]

Bormann and Likens had stressed that the ecosystem approach was uniquely able to take account of all significant aspects of the landscape. To fulfill this in their own study they recruited students and colleagues to research other aspects of the terrestrial ecosystem, including its streams and fauna. In June 1968, for example, Stuart Fisher, one of Likens's few doctoral students at Dartmouth, began a study of Bear Brook, a stream draining an undisturbed watershed. He developed an energy budget for the stream, showing its import, export, and accumulation of organic matter.[80]

Other researchers studied mice, salamanders, and deer. The largest effort, however, was devoted to birds. Shortly before Likens left Dartmouth for Cornell he persuaded two ornithologists, Frank Sturges and Richard Holmes, to work at Hubbard Brook. As had the ecologists studying other animals, they began by examining the role of birds in nutrient cycling and energy flow. This was, Bormann and Likens explained, to help fill gaps in their understanding of ecosystem processes. "Another consideration in determining the complete balance of these forested watersheds," they noted in justifying these studies, "is the withdrawal and recycling of various nutrients by terrestrial animals and the community of plants and animals associated with the small drainage streams."[81]

These studies concluded that, at least for these ecosystem processes, birds, like other animals, were insignificant. They had only a tiny role in the movement or export of nutrients. For Sturges and Holmes, interesting research questions clearly lay elsewhere, and they shifted away from the role of birds in ecosystems to studies of their behavioral ecology, community structure, and population dynamics. Holmes also believed such studies to be more amenable to experimental study, and therefore more likely to be funded and to be published in leading journals.[82]

But by following these questions of interest to population and community ecologists, Holmes and Sturges diverged from the concern for ecosystem structure and functioning that characterized the Hubbard Brook study. This divergence also occurred in other aspects of the Hubbard Brook study. Stream studies, for example, focused less on the place of the stream in the watershed ecosystem than on decomposition processes, invertebrate community structure, and other internal aspects of the stream ecosystem.[83] Such research nevertheless remained at Hubbard Brook because its large area, protection from disturbance, and availability of long-term records of environmental changes were still useful for their research.[84]

This divergence reflected, in part, the division between population and ecosystem studies characteristic of the discipline of ecology. But there were also several factors specific to Hubbard Brook that can account for this pattern. As the study expanded, it involved researchers from more institutions. Holmes, for example, came to Dartmouth after Bormann had left and remained when Likens moved to Cornell. This geographic separation prevented the close collaboration that Bormann and Likens had found necessary to a common ecosystem approach. A likely second factor was the involvement of students. In order to conform to the constraints of time and resources of a doctoral program, students often limited their attention to a single aspect of the ecosystem. Some students, such as Marks, did succeed in developing the implications of their research for the entire ecosystem. But this also reflected Marks's close collaboration with Bormann. As the study expanded, such close collaboration may have become less possible.

Another important factor was the management strategy of Bormann and Likens. While molding research into directions that could contribute to the overall program, they also protected the independence of researchers. No participants, including students, have been required to study a problem they had not developed themselves, they claimed. They repeatedly stressed the significance of this: "From the beginning we have called attention to the kinds of information a cooperative study might produce, but always we have encouraged individuality in the design and execution of research. *We deem this individual research freedom one of the greatest assets of the Hubbard Brook Study.*"[85] Both strongly believed that scientists would do their best work if they defined their own problems. This view, of course, recalls similar values expressed by ecologists at the Nature Conservancy and at Oak Ridge from the 1950s through to the International Biological Program.

As ecologists and students came to Hubbard Brook to study only specific aspects of the watershed, integration of results into an overall theory of ecosystem behavior became the special responsibility of Bormann and Likens. By 1974 they and a few of their colleagues had assembled initial results and explained their significance for ecosystem theory. The influence of their presentations was beginning to be felt within the research community, as ecologists elsewhere adopted the watershed technique to study ecosystems.[86] But by this time they had also decided that only within a book could they fully develop the implications of their results. Their first, *Biogeochemistry of a Forested Ecosystem,* appeared in 1977. It summarized their understanding of the import, export, and cycling of nutrients within forest ecosystems, explaining the significance of their data on precipitation and stream-water chemistry, weathering, and the movement of water. It was, in effect, a demonstration that they had achieved a central objective of their study, the inclusion of physical and chemical phenomena into ecological research, using the small watershed technique.

This was also one of the objectives of their second and more ambitious book, *Pattern and Process in a Forested Ecosystem,* which appeared in 1979. Bormann and Likens considered the book to be their most successful presentation of an integrated ecosystem perspective.[87] At least five years in the writing, it combined biogeochemical insights with their understanding of the temporal development of the northern hardwood forest ecosystem. In writing it they sought to demonstrate that by integrating the results of Hubbard Brook research they could approach, in Alex Watt's words, the "idealistic objective of fusing the shattered fragments into the original unity."[88]

The foundation of their theory was a conceptual model of forest ecosystem development, the Biomass Accumulation Model. The forest, they explained, passes through four phases. Immediately after a major disturbance, such as a clearcut, it enters a "reorganization" phase lasting ten to twenty years. That stage is characterized by rapid accumulation of living vegetation but an even greater loss of dead biomass, in part through the nitrification-induced loss of dissolved nutrients. Eventually, however, the growth of trees and other vegetation would outweigh this loss of biomass and the ecosystem would enter its "aggradation" phase, marked by a continual increase in biomass as the forest becomes dominated by a few large trees, all of the same age. After more than a century these trees would begin to be vulnerable to windthrow, insect pests, or other hazards, and would one by one topple over and be replaced by smaller, younger plants. As more large trees fell, their loss for a time outweighing the growth of the smaller plants replacing them, forest biomass would begin to decline. This period of decline was the "transition" phase, of no set length. Eventually the forest will be composed of trees of all ages, of a total biomass fluctuating around a mean less than the maximum achieved in the aggradation phase. This would be the Shifting-Mosaic Steady State phase, which would continue until the next clearcut or other major disturbance. In describing this steady-state phase as a mosaic of trees of varying ages, Bormann and Likens recalled Watt's similar analysis three decades before of the dynamics of plant communities.

While these phases were defined in terms of biomass, they could also be distinguished by the ecosystem's stability—that is, its ability to control the destabilizing forces (such as wind or water) affecting it. A stable ecosystem would be able to minimize its loss of water and nutrients and to control erosion in the face of these forces. By these criteria, the aggradation phase was the most stable, the reorganization phase the least stable; the transition and steady-state phases were of intermediate stability.[89] The varying stability of these phases reflected the significance of plant biomass. Through transpiration, interception of raindrops, nutrient uptake, and a variety of other mechanisms, vegetation played a pivotal role in retaining water and nutrients within the ecosystem.

A chief aim of Bormann and Likens was to demonstrate the value of studying the abiotic as well as the biotic components of the ecosystem, and the incompleteness of ecosystem theory not based on this approach. In introducing their book they set themselves in opposition to those ecologists who did not share their view of the vital necessity of understanding the inanimate processes of ecosystems.[90] The result of this incomplete approach, they argued, was needless misinterpretation. For example, the forest ecosystem was widely considered to be most unstable immediately after a clearcut, as indicated by the heavy loss of nutrients. But this loss in fact forestalled even greater losses. The released nutrients "irrigated and fertilized" the ecosystem, permitting more rapid recovery of vegetation, thereby preventing excessive erosion. "In essence," they explained, "the ecosystem draws on a bank account of energy and nutrients built up over a long period of time to solve an immediate crisis that threatens still-greater capital losses."[91]

Conversely, the final phase of ecosystem development, widely considered the most stable (as, for example, in Eugene Odum's well-known hypothesis of ecosystem development), was actually, Bormann and Likens argued, less stable than the earlier aggradation phase. The problem, they insisted, was that many ecologists did not adequately consider the complexity of ecosystem development. This complexity had, of course, been for several years a special focus of research at Hubbard Brook. This was, in particular, a chief shortcoming of those ecologists presenting competing explanations of the behavior of watershed ecosystems:

> In most discussions of steady state or "climax" ecosystems (e.g., Vitousek and Reiners, 1975), the condition of the steady-state ecosystem is loosely defined by one or two whole-ecosystem parameters, such as biomass accumulation equals zero or gross primary productivity equals total-ecosystem respiration. More often than not, authors are extremely vague or totally unconcerned about the fine structure of the steady-state ecosystem or the means by which the steady state might perpetuate itself through time. Yet such information is basic to any analysis of the energetic, biogeochemical, or stability relationships of the steady-state condition.[92]

It was this "fine structure" of ecosystem development, as well as the results of the JABOWA simulation, that convinced them that the model accepted by many ecologists, in which biomass increases asymptotically to a maximum in the steady-state climax, was not correct, at least in the case of the northern hardwood forest.[93]

The perspectives developed in these two books integrated results from dozens of research projects completed at Hubbard Brook during the previous fifteen years. From JABOWA (the simulation model of forest development), Bormann and Likens took predictions of biomass accumulation over several hundred years of forest development. From separate studies by Dominski, W. W. Covington, and Gosz, they took data on the changing character of the forest floor during succession. From Marks's studies, they took results on the recovery of vegetation from clearcutting. From the research of Pierce, Hornbeck, and Federer they took information concerning the role of plant evapotranspiration and other mechanisms in watershed hydrology. From numerous studies they derived conclusions relating to the role of vegetation in the regulation of export of nutrients from the watershed. For each Hubbard Brook study they mentioned, they explained its link to their overall theory of forest ecosystem behavior.

Thus, the integration of Hubbard Brook research into ecosystem theory was the special responsibility of Bormann and Likens. This has implications for our understanding of the organization of large-scale science. In terms of scale alone, the Hubbard Brook study is an example of big science: it represents the combined effort of many scientists working within a single project. However, as Joel Hagen has observed, it has not adhered to the hierarchical, vertically integrated corporate or military model of organization described as characteristic of the International Biological Program or as epitomized by research using particle accelerators, space telescopes, and other large-scale hardware.[94] Instead, it illustrates two other ways, in addition to vertical integration, in which big science may be organized. One is horizontal integration: a decentralized network of scientists and students dispersed across the watershed, communicating with each other informally as colleagues or advisers, with their results collected and synthesized by Bormann and Likens. The other is temporal integration: the conduct of a long-term experiment by a series of researchers, each of whom may be present only for part of the experiment but whose results are combined into a synthetic understanding of the phenomenon studied. The study of ecosystem disturbance and recovery at Hubbard Brook is one example of this form of integration.[95]

Bormann and Likens's role in integrating the results of the study also had specific implications for the links between the study and environmental issues. Bormann and Likens believed that ecological knowledge would be most relevant to these concerns when expressed in terms of the overall ecosystem. In fact, they considered the synthesis of ecosystem theory, and the packaging of research results into a format useful to environmental managers, to be largely synonymous. As they had written in introducing their first book, their goal was "to order some of this information in ways we consider useful both to the scientists concerned with the theory of biogeochemical cycles and the structure, function, and development of forested ecosystems, and to the land-use specialists who are concerned with the production of a variety of benefits, goods, and services from northern hardwood ecosystems."[96] Thus, the task of integrating results became equivalent to that of developing their environmental implications. In contrast, most other Hubbard Brook researchers, who studied only specific aspects of the watersheds, rarely noted these implications.

Bormann and Likens accordingly concluded *Pattern and Process in the Forested Ecosystem* with a lengthy discussion of the practical implications of their work. Some of the implications were fairly specific. The behavior of nutrients during the reorganization and aggradation phases of ecosystem development, for example, indicated that their loss from the forest in the years after cutting could, contrary to the belief of some foresters, far outweigh the amount removed in wood products. This had to be considered in calculating the date of the next cut to ensure that there was sufficient time to replenish lost nutrients. The time between harvests, they suggested, should be at least sixty-five years. Harvesting techniques that disrupt the forest floor, such as removal of the roots, could also have a large impact on the loss of nutrients, and should be discouraged. They made a variety of specific recommendations regarding choice of sites for harvesting, size of cuts, design of roads, protection of streams, and protection of species of no economic value (such as pin cherry) but significant in ecosystem recovery.[97]

Behind these recommendations lay their belief that only by regarding the forest as an ecosystem could foresters consider all the ecological aspects of forest harvesting, including those, such as nutrient depletion and impacts on marginal species, that were usually neglected. Studies of ecosystem processes also suggested that forests provide a range of benefits to humans—besides timber. These included regulation of streamflow and erosion, moderating the local climate, and replenishing water supplies. Most important, forests regulate the chemical quality of the water percolating through them. They absorb much of the matter brought to them by the atmosphere, including such pollutants as sulfur dioxide, ozone, and heavy metals in both air and water.[98] Bormann and Likens stressed that the values of forests extended beyond those addressed by the short-term considerations of marketing and profitability. Forest management, they believed, needed to be guided by more than markets. Rational, long-term management based on scientific principles was required: "A longer planning frame would allow areas utilized for commercial forestry to serve more effectively a multiplicity of uses and needs, including those of interest to forest industrialists, conservationists, and recreational and regional planners. These benefits result from a planned long-term harvesting schedule rather than a boom-or-bust schedule governed solely by market conditions."[99]

This argument—that ecosystem research could provide the basis for a more comprehensive, rational approach to environmental protection and management—was frequently made by ecosystem ecologists. By stressing the practical value of the research, as well as the necessity of long-term study independent of immediate practical concerns, Bormann and Likens, like many of their colleagues, asserted the need for independence for their research in order for it to be of greatest value to society.

Bormann and Likens also took pains to point out that much of their research results had been anticipated by foresters. Just as Covington and Dominski had found that the forest floor continues to decrease in depth for several years after clearcutting, so had foresters. Monitoring of the deforested watershed had shown that erosion did not begin until after nearly two years; this confirmed the "findings of foresters . . . that the removal of forest cover . . . has little effect on sediment yield as long as the forest floor is intact." Their hypothesis of the likely appearance of the steady-state phase was, they noted, much like the description of "virgin" New Hampshire forest recorded by a forester three-quarters of a century before. Indeed, much of their results confirmed, as they noted in reference to their studies of commercially clearcut areas, "what foresters have known for some time."[100]

By noting that their results were consistent with those obtained by foresters, Bormann and Likens reinforced their long-standing contention that an essential feature of Hubbard Brook research was its infusion of insights from forestry into ecology. By linking their results with those obtained by foresters they enhanced their credibility, particularly among foresters. But they went beyond this by providing quantitative assessment of the more qualitative observations made by foresters. Thus, while foresters measured "sediment" eroded from watersheds, the ecologists measured the elemental content of this sediment. Similarly, while foresters had measured with a ruler the thinning of the forest floor after logging, the ecologists used sophisticated techniques of nutrient analysis.

The books written at Hubbard Brook and the Oak Ridge ecosystem simulation models represented contrasting approaches to the integration and communication of ecological understanding.[101] Both served to combine the results of studies of aspects of ecosystems to generate a synthetic understanding. But their purposes, and place within the research process, also differed somewhat: simulation models served primarily as research tools, identifying for Eastern Deciduous Forest Biome ecologists knowledge gaps to be filled by further studies; Hubbard Brook books, in contrast, provided a comprehensive statement of results and interpretation as the capstone of fifteen years of research. As communication tools the books were more successful, among both ecologists and those interested in the environmental implications of ecosystem ecology. The models were neither intelligible to most outsiders nor specific enough for most practical applications. The books, in contrast, provided an explanation of forest ecosystems that was accessible to intelligent nonscientists and useful to ecologists and other scientists, including foresters.

During the 1970s Bormann and Likens increasingly reached beyond the ecological research community, to both the forestry profession and the general public. Working closely with Pierce and his colleagues, they helped transmit ecological insights to foresters. In the late 1970s the merits of various methods of forest harvesting—selective cutting, clearcutting, stem-only harvesting, whole-tree harvesting, and complete tree harvesting—were being debated in New England and elsewhere. Bormann, Likens, and their colleagues aimed to contribute to this debate. One way was by organizing a seminar for industrial foresters to explain what they had learned at Hubbard Brook.[102] They also began a new watershed experiment to evaluate the ecological impact of whole-tree harvesting. In addition, they wrote for the general public, explaining the noneconomic benefits provided by forests and the risk posed by ecologically unsound forest management and air pollution.[103] By the late 1970s they probably considered this translation of their results into terms important to the public to be less a risk to their professional credibility than it had been during the age of ecology that had begun the decade.

These efforts to derive practical implications from their research and to communicate them to those who could apply them were not aimed at ensuring continued funding. Unlike those agencies with environmental protection or management responsibilities, such as the Environmental Protection Agency, the National Science Foundation was at best indifferent to practical applications. With only one exception (an experimental study of stream acidification, funded by the EPA), Bormann and Likens did not seek funding from mission-oriented agencies. Like many, they believed that such agencies interfered too much in the research they supported.[104] Again, there is an evident contrast with Oak Ridge, which received support from an increasing variety of sources during the 1970s.

But there were other considerations motivating Bormann and Likens to make this effort. One was their personal concern for the protection of the forest environment. Another was their conviction that publicly funded researchers had an obligation to return some benefits to society. This was especially important, they believed, for a relatively expensive project such as Hubbard Brook, which during the 1970s received about $200,000 per year from the NSF, as well as much indirect support from the Forest Service, universities, and other agencies. And finally, by demonstrating a capacity for deriving recommendations for forestry management, Bormann and Likens helped support their contention that the Hubbard Brook study had strong links with forestry research and that it was in part through these links that it could make unique contributions to ecosystem ecology. Having produced such recommendations, Bormann and Likens's use of the methods of foresters, and their use of forestry research as independent confirmation of their results, appeared more credible.

Bormann, Likens, and the other ecologists at Hubbard Brook derived valuable benefits from their collaboration with foresters. They had the use of protected research sites equipped with weirs and weather stations. With this equipment they could develop a distinctive approach to ecology that recognized the significance of both the biotic and abiotic fractions of forest ecosystems. This research was built on the foundation of an understanding of the interaction between nutrient cycling and the movement of water into, within, and out of watersheds. Much of this foundation was itself fashioned from the earlier insights of forestry researchers. Some forestry research, including study of the effects of actual cutting, also helped corroborate their experimental results. At the same time, they were able to demonstrate their conception of the value of ecosystem research to environmental concerns. Their theoretical understanding of ecosystem processes, while often not immediately applicable to specific forest management problems, could provide a framework within which forest research and management could be integrated.

This close cooperation between foresters and ecologists was possible only because of the presence of scientists familiar with both disciplines who could bridge the gap between them. Bormann, whose education had included both, was the first bridge. Likens, who quickly developed his experience in forest ecology to match his knowledge of limnology and biogeochemistry, was another. When the need arose to test the relevance of the results of the first experiment to forest management, Pierce and his colleagues adopted some of the practices of the ecologists. By 1970 they too were acting as bridges between ecology and forestry.

The structure of the Hubbard Brook study, as a cooperative endeavor of academic ecologists and forest service researchers, reflected the bridge-building efforts of its chief participants. That they did not separate reflected the commitment of the individuals to their alliance. This commitment can in large part be traced to the awareness of Bormann and Likens of the unique benefits they derived from their cooperation with foresters in furthering the innovative watershed approach that was Hubbard Brook's distinctive contribution to ecosystem research.

One of the singular aspects of the Hubbard Brook study was the effort to draw on forestry research to generate a new approach to ecology while using this approach to revise forestry practices. I continue this theme of interaction between natural resource management and ecology in the next chapter. I recount an episode in the history of ecology in Canada to illustrate how certain conditions and scientific perspectives led to ecology playing a supportive, legitimating role in natural resource management.

PART THREE

Canada

7

Ecology and the Ontario Fisheries

The Great Lakes must have appeared almost unlimited to those who were first to gaze upon them: inland seas extending deep into North America, ready, with their surrounding landscape, to provide seemingly infinite water, wood, furs, and cargo-carrying capacity. Not least was their promise as a source of food. Whitefish, lake herring, lake trout, sturgeon, salmon, and more than 150 other species of fish were staples in native communities. And to the first European settlers the lakes appeared as fertile as the surrounding land. In 1835 one observer wrote of Great Lakes fish, "Their quantities are surprising, and apparently so inexhaustible as to warrant the belief that were a population of millions to inhabit the lake shores they would furnish ample supplies of this article of food without any sensible diminution."[1]

As the first of these millions settled in the region, fishermen worked mainly to satisfy local demand, with a small surplus salted and transported outside the region. Their take increased as the market grew and as new technology became available: improved nets, fish freezers, and steam-powered fishing tugs. By the 1870s fishermen were harvesting the fish in a fashion typical of natural

resources exploitation during this period: "Astounded at the seemingly inexhaustible bounties of nature, [their] one aim was exploitation in such a man-
ner as to reap the largest possible rewards in the shortest possible time."[2] By
1900 the Great Lakes commercial fisheries employed 10,000 people. Fish
stocks declined under this pressure, and some species disappeared forever.
Other human interventions—including domestic, industrial and agricultural
pollution, accidental or deliberate introduction of exotic species, and elimination of wetlands and other fish habitats—resulted in further damage to fish
populations.

The growth of commercial fisheries on the Great Lakes was eventually
accompanied by the imposition of fisheries management, as well as scientific
research on fish populations and their environment.[3] The provincial government in Ontario had primary responsibility for Canadian fisheries management on the Great Lakes. In 1946 it and the University of Toronto entered
into a partnership to conduct research in freshwater ecology and fisheries.
During the next twenty-five years the university strengthened its status as a
major academic center for ecology, and the largest provincial program of
fisheries research in Canada provided the basis for fisheries management
practices in the Canadian portion of the Great Lakes, as well as in smaller
lakes in the province.

This arrangement between the Ontario government and the University of
Toronto resembled in some respects the organization of the Hubbard Brook
Ecosystem Study. In both cases, and in contrast with the Nature Conservancy
and Oak Ridge National Laboratory, academic researchers worked in cooperation with a natural resource management agency. But the Ontario arrangement differed from Hubbard Brook in several respects. Most obviously, the
concern was fish, not forests. The issues of immediate practical concern also
varied: the maintenance of fish stocks in the face of a variety of human impacts, not the ecological implications of forest harvesting. A third difference
was in the relative importance of university researchers to the resource management agency. For two decades scientists at the University of Toronto
strongly influenced both the research and the management activities of the
Ontario government agency, while Hubbard Brook ecologists were relatively
much less important to the U.S. Forest Service, which had its own research
and managerial perspectives.

By the early 1970s the context of ecology was being transformed: the Great
Lakes fisheries were in serious decline, and changes were occurring in the political landscape, in society's interest in the aquatic environment, and in the
research interests of university ecologists and fisheries scientists. Contemporaneous transformations in context had led to changes in research practice at

Hubbard Brook and Oak Ridge, and the same thing happened in Ontario. The new approach to research was marked by a broader perspective on fish communities and their environment, as well as by a new plan for fisheries management, the "Strategic Plan for Ontario Fisheries," that emphasized protection or rehabilitation of fish communities. These shifts were accompanied by changes in relations between the provincial government and the university as the priorities and requirements of both resource managers and scientists evolved.

ORIGINS OF A PARTNERSHIP

In the summer of 1921 Professor W. A. Clemens of the University of Toronto led a small party in a study of fish populations in Lake Nipigon, in northwestern Ontario. This expedition was the first major project of the Ontario Fisheries Research Laboratory (OFRL), established the previous year within the university's department of biology.[4] B. A. Bensley, at the time chair of the department, set out in 1922 the objectives of the OFRL, and its scientific and economic justification. It could, he explained, contribute to a better understanding of lakes as "complete physical-biological complexes" or as "small and complete world[s]" characterized by a "natural balance" between organisms, as had been perceived by Stephen Forbes and other American limnologists. Through study of the distribution, movement, growth, prey, predators, and parasites of important commercial and game fishes, it could also provide a scientific basis for economically sound fisheries regulations and management.[5] Thus, the OFRL would contribute to scientific objectives while meriting support from a provincial government concerned with the economic benefits of natural resource development. There was logic also in seeking research support from the province rather than the federal government. The latter had abolished the Commission of Conservation (its most significant initiative in resource management) the previous year, and the federal Biological Board of Canada had focused its attention on the Atlantic and Pacific fisheries. Thus, the provincial government was clearly more likely to support studies of inland fisheries.[6]

University biologists led by William Harkness, director of the OFRL from 1924 to 1946, continued to emphasize in their writings the economic relevance of the laboratory, and therefore the appropriateness of provincial support. Fisheries management, they argued, should be based on a scientific understanding of fish populations, and the factors that determine the productive capacity of lakes and streams. Echoing the "wise use" sentiments of Gifford Pinchot and other leaders of the progressive conservation movement,

Harkness stressed the importance of research to fisheries conservation: "One of the immediate aims of the laboratory has been to show the way for conservation by which this is meant the full but proper utilization of our water resources. Conservation is based on knowledge; knowledge of the complete life history, food requirements, and optimum living conditions of the organism or organisms under consideration; knowledge which must be possessed by those in administrative capacity and by all others who are casually or constantly in contact with the organism either in the capacity of protecting it or destroying it for sport or commercial purposes."[7]

Such arguments had some success. The Department of Game and Fisheries provided grants to the OFRL and in 1925 established a biological section to conduct lake surveys, which it expanded in 1928 into a Biological and Fish Culture Branch. The Ontario Department of Lands and Forests (ODLF) also assisted, providing facilities and equipment in Algonquin Park. This made possible an intermittent program of fisheries and a limnological study of Lake Erie, Lake Ontario, and several smaller lakes, conducted by Clemens, Harkness, John Dymond, Fred Fry, and their students.[8] Much of the research, beginning in 1936, was based at a small joint provincial-university lab on Lake Opeongo, in Algonquin Park. These studies focused not only on fish populations but also on other biota, and physical and chemical aspects of lakes. Among various specific projects were some of the earliest studies of the ecological impact of DDT, conducted in 1944 and 1945.[9]

During the 1920s and 1930s University of Toronto scientists were clearly more successful than their ecologist counterparts in Britain in gaining government support. One reason was that in Canada there existed an emerging tradition of fisheries research, fostered by the Biological Board of Canada (of which much of the leadership was provided by Toronto scientists); arguments concerning the practical utility of such studies therefore already had credibility. Also, British ecologists had to seek support in this era from forestry or agricultural agencies, both of which were reluctant, as Tansley and his colleagues had found, to support a separate discipline. But in Canada, fisheries science was the only discipline relevant to aquatic resources. Once the case had been made that management of these resources required research, fisheries science had no competition for the funding that could result.

CONSERVATION AND FISHERIES SCIENCE

Immediately after the Second World War, an opportunity appeared to strengthen the relationship between the university and the provincial government. Since the 1930s, naturalist, forestry, and agricultural groups in Ontario had been drawing attention to the erosion and flooding affecting Southern Ontario farmlands and towns. Such problems, they argued, demonstrated a need for comprehensive, scientific conservation of land and water resources. Examples of conservation work elsewhere, including the Civilian Conservation Corps in the United States, reinforced their arguments. In 1941, advocates of conservation stepped up their activities, with scientists from the University of Toronto, including Dymond and Alan F. Coventry, and from other provincial universities and colleges providing much of the leadership. By organizing meetings and writing booklets and articles on the need for conservation, they firmly established a link between conservation and scientific expertise.[10]

This advocacy resembled the contemporaneous wartime initiatives of British ecologists. In both cases, conservation was portrayed as a scientific activity requiring expert advice and government support.[11] But they also differed: while British scientists focused on the study and management of nature reserves, their Canadian counterparts emphasized erosion, deforestation, flooding, and other problems impeding full and sustainable use of land and water resources. This difference reflected the participation of different disciplines. While ecologists had the central role in British advocacy of conservation, conservation in Ontario was advocated by scientists from several disciplines and professions, including fisheries science, wildlife management, soil science and forestry, as well as ecology. In contrast to Britain, where there was a cohesive if underfunded discipline of ecology led by Tansley, Elton, and others possessing a strong sense of the past and potential future of their discipline, in Canada the ecologists had little common identity and no national society, journal, or dedicated sources of funding with which to build this identity. Instead, those who did ecological research usually associated themselves with natural resources management—as, for example, University of Toronto scientists did with fisheries management.

After the war the Ontario government, like the British government, proved receptive to conservation. Conservation promised both a more effective approach to resource management, and employment for returning veterans. Accordingly, it began several conservation initiatives. These were not associated with ecology, as the Nature Conservancy had been, but with a broader, natural resources management perspective. One initiative, begun in 1946, was a reshaping of its approach to fisheries management. Its Department of Game and Fisheries had considered itself primarily a law enforcement agency, with game wardens controlling poaching and overfishing in order to ensure some fairness in the distribution of the take among fishermen and to reduce conflict among them. That year, however, this department was amalgamated into the

Department of Lands and Forests, becoming its Division of Fish and Wildlife. Confirming that management was to be based on science, Harkness, director of the OFRL, became director of the division.

It was immediately apparent that the Department of Lands and Forests was more hospitable to research than the old Department of Game and Fisheries had been. One year after the amalgamation, the chairman of the provincial Advisory Committee on Fisheries and Wildlife noted with satisfaction that "under the [ODLF] the necessity for research as a basis for the administration of the fisheries and wildlife resources of the Province is realized and as a consequence many research programs have been initiated under the auspices of the Department."[12] This view of science as integral to fisheries management was espoused not only by fisheries scientists. R. N. Johnston, a forester who was director of the ODLF's Research Division, noted in 1951 that "our working assumption . . . has been that fisheries management will have a very dim future indeed unless research is vigorously organized and driven to produce the many necessary facts and improved techniques and procedure on which sound and increasingly productive management must be based. In other words, no successful long term research, no successful long term management."[13]

The division quickly built up a corps of scientific expertise. It had begun in 1946 with two biologists; by 1952, it had sixteen, and by 1964, it had sixty-seven biologists. During this period a small network of fisheries research laboratories was also established. The ODLF assumed support of the Opeongo Laboratory in 1946. The same year a limnological laboratory was established at the department's research station at Maple, north of Toronto. The following year a research station was established at South Bay, Manitoulin Island, on Lake Huron. These were followed by a station at Wheatley, on Lake Erie, in 1953, and another at Glenora on Lake Ontario, in 1957.

At the center of this expansion was the University of Toronto. The Opeongo and Maple laboratories, by formal agreement, were managed jointly by university and government scientists, and provided facilities for many graduate students. Fisheries research elsewhere was also shaped by university scientists. For its first few years, Fred Fry directed operations at the South Bay laboratory, and the Glenora laboratory began by taking over a study of lake whitefish that Fry had initiated.[14] University scientists also influenced research through their membership on the provincial Advisory Committee on Fisheries and Wildlife Research, and conducted occasional special studies for the department.[15] Fry and other university scientists trained numerous provincial researchers and fisheries managers.

By the late 1940s, a close, mutually beneficial relationship had developed

between university and government. The university received support and facilities for research and for teaching advanced students. Stable, long-term
government support for collection of statistics of fishery catches, effort, and
sales was especially important. In return, the government received advice, a
steady supply of trained biologists, and a scientific basis for and justification of
management activities. The key role played by University of Toronto scientists
in fisheries management reflected two conditions specific to the Canadian
context. One was the provincial governments' jurisdiction over natural resources, including inland fisheries. This meant that the Ontario government
had chief responsibility for both fisheries management and related research.
The federal government's involvement in inland fisheries research—on such
topics as the sea lamprey and Lake Superior—and the support provided by
the National Research Council was less significant. The other condition was
the close connection between the university and the provincial government.
The University of Toronto was the most important element of the provincial
university system and possessed numerous formal and informal ties with the
provincial government, as well as a location across the street from the legislature. So provincial officials had become accustomed to calling on the University of Toronto for scientific advice.

Thus, as I have noted, the University of Toronto–Provincial Department of
Lands and Forests relationship resembled that between university scientists
and the U.S. Forest Service at Hubbard Brook. Both are examples of academic
scientists and resource management agencies developing mutually beneficial
ties of cooperation. However, while Hubbard Brook ecological research did
not have a direct impact on Forest Service policies, University of Toronto scientists played a key role in defining ODLF research policy and, indirectly,
fisheries management policy. This was, in part, because while ecology at Hubbard Brook drew from forestry, it had a separate identity and differing priorities, research topics, and the like; ecology at Toronto, in contrast, was more
closely associated with fisheries management. This reflected the history of association between natural resources management and ecological research in
Canada.

From the late 1940s to the 1960s, university and government researchers
sought to determine the size of Great Lakes fish populations, the causes of
fluctuations in these populations, the impact that fishing and other factors had
on these populations, and the relation between fishing intensities and yields.
Data on fish populations, including catch size in relation to fishing effort, size
and growth rates, sex, fecundity, and age structure, was obtained through surveys of commercial and sportfishers and from experimental fisheries. At the
South Bay station, for example, one experiment imposed equal fishing pres-

sure on "worthless" and sought-after fish species to determine whether this would result in a higher yield of the more desirable species.[16] Results from this and other projects then contributed to both specific management issues, such as the determination of the impact of the lamprey on fish populations in the upper Great Lakes, and to the overall objectives of predicting future fish populations and estimating the level of exploitation that would ensure a maximum yield. Researchers generally focused on the dynamics of fish species of importance to commercial or sportfishers.

Some of the most important fish populations in the Great Lakes fluctuated dramatically from year to year. Accordingly, determining the cause of these fluctuations, and eventually predicting them, became a research priority. This research was guided by certain assumptions. According to Harkness, Fry, Dymond, and other biologists, the number of fish produced each year was determined largely by environmental conditions. If the lake or river was a suitable habitat for reproduction, then a large fish population would surely result. On the other hand, the number of adult fish had little to do with the number of offspring produced each year. Given that a single female could produce several tens of thousands of eggs, they reasoned that even only a few adult fish could repopulate an entire lake. As Fry explained in 1964, "Because we have these large differences in the production of young in different years we know that it isn't just the amount of spawning stock that brings success to a fishery. Indeed we are coming to quite the contrary opinion. We are beginning to believe a favourable environment is much more important than a super-abundance of spawning fish."[17] This reasoning helped stimulate the study of fish populations in relation to changes in the environment. Annual fluctuations in Great Lakes fish populations must reflect the impact of environmental conditions on the survival of fish during the most vulnerable stages of their life history. It was, in particular, considered likely that a correlation between summer temperature and size of year-classes in fish populations could be identified. Accordingly, several studies at the stations on Lake Ontario, Lake Erie, and Lake Huron tried to relate changes in fish populations to water temperature, as reflected by meteorological records.[18]

The view that the physical environment is most important in determining the size of each species' year-class was reflected in skepticism toward fish culture. While it had been widely assumed, prior to 1950, that culture (the artificial rearing and release of fry) could supplement or replace natural reproduction, a study testing this assumption had begun within the OFRL by 1944. Young fish were cultured and released into Lake Ontario in alternate years, and fishery statistics were then examined to determine the effect of this stocking on the age structure of fish populations.[19] While the study demonstrated that fish culture made an insignificant contribution, such a conclusion was already implicit in the assumption that Great Lakes fish populations were limited by the environment, not by the availability of spawn. As Dymond (fig. 7) commented in 1964, "The principal reason why the planting of artificially reared young does not materially increase production is . . . that even when the population of spawners is low, there are usually enough to produce an abundant crop if conditions are satisfactory."[20]

The relation between environmental factors and fish was of particular interest to Fry (fig. 8). Beginning in the mid-1940s he applied his wartime experience in the study of human physiology, and his understanding of the "limiting factor" concept of the British botanist F. F. Blackman (which had also become more widely used by ecologists, as in, for example, Charles Elton's 1927 text *Animal Ecology*) to the experimental study of the interaction between factors such as temperature and dissolved oxygen on fish physiology and activity.[21] He established the Maple laboratory as a center for "experimental limnology," by which he meant laboratory study of the physiological

Fig. 7. John Dymond (courtesy of department of zoology, University of Toronto)

Fig. 8. Fred Fry (courtesy of J. S. Griffiths, Ontario Hydro)

responses of fish to their physical and chemical environment. There, he and a succession of students pursued experiments on such topics as the temperature tolerance and acclimation potential of fish species, the determination of lethal and optimum levels of oxygen for fish and their eggs, and their consumption of oxygen at various levels of activity. Studies were also pursued on other topics, such as water chemistry.[22]

Research at Maple was often closely integrated with field studies at the Great Lakes stations, with results in one environment suggesting new problems to explore in the other. For example, if in the laboratory a preferred temperature was estimated for a certain fish species, then field study could be undertaken to see if that species did in fact occur at such a temperature. Conversely, indications from fishery statistics that yields were affected by certain environmental conditions could be tested in the controlled conditions of the laboratory. Both field and laboratory studies, therefore, while consistent with the widely held assumption that environmental conditions determined Great Lakes fish populations, also contributed to Fry's conceptual framework integrating fish physiology, population dynamics, and environmental conditions.

In the field, as in the laboratory, experimental methods of studying the fishery assumed central importance. By the late 1940s, experiments encompassing entire lakes, or part of one of the Great Lakes (in the case of the South Bay research station) were underway. In one study, nutrients were added to four of the Algonquin Lakes to see whether that would lead to greater fish production. In another, trout lakes in the park were closed to fishing in alternate years, to see whether fish yield would increase.[23] In a third study, as I have noted, an experimental fishery at South Bay examined the effects of equal pressure on all species of fish.

In the eyes of university researchers, these experiments had several advantages. As "experimental management" they provided a bridge between research activity and management, making it possible to test management practices under more controlled conditions, but on a realistic scale. It was analogous, in their view, to experimental agriculture.[24] These experiments also provided opportunities for studies not directly related to fisheries.[25] More generally, as attempts to apply experimental methods to field conditions, these studies had obvious parallels with forest plot studies and experimental management of nature reserves in Britain, field studies of radiotracers at Oak Ridge, and watershed studies at Hubbard Brook.

In discussing their work and its relation to fisheries management, University of Toronto fisheries researchers commonly set ambitious goals while warning that achieving them would require patience. One such goal was to manage, even control, Great Lakes fisheries in order to produce the types and quantities of fish demanded. Often, they made an analogy with agriculture. As Dymond argued, "Although fisheries research has not yet made it possible to manage the fisheries as effectively as agricultural crops can be managed, it is not to be assumed that such research is not making progress towards its ultimate goal of producing at will the kinds and numbers of fish desired. The slowness of progress is due to the complexity, and other features, of water conditions. It is only through the slow accumulation of knowledge by research that the ultimate objective will be reached."[26] Similarly, R. R. Langford noted that "in these days of such rapid scientific progress we should be able to look forward to truly remarkable improvement in the methods of controlling both the quantity and quality of the products of lakes, ponds, streams and perhaps even the oceans."[27] The Department of Lands and Forests reiterated this goal and the agricultural analogy, noting that the ultimate objective of research is "to produce maximum and continuing crops of the desirable species and to increase the use of presently undesirable ones when possible."[28] These views reflected an important aspect of fisheries management between the 1940s and 1960s: a consensus on its central goal of achieving the maximum possible production of fish within the constraints imposed by Ontario's aquatic environment. In viewing this resource as a "crop" subject to manipulation, this outlook was also entirely consistent with the dominant ideology of natural resource management, as it had originated in the progressive conservation movement of the turn of the century and had been reiterated in such contemporary texts as Aldo Leopold's *Game Management*.

But this goal, as Dymond and other researchers stressed, demanded long-term, systematic study, including the assembly of fishery statistics covering years, even decades, before definite patterns in the dynamics of populations could be identified, let alone controlled. In the meantime, much of the research, while directed toward the distant objective of predicting and achieving maximum yield, was not immediately relevant to practical fisheries problems. This reflected the latitude that scientists had in defining their research priorities; as Fry explained in 1956, "We are employed in the summers as individuals by the government and are given great freedom by that agency in furthering our research."[29] They also defended this latitude in terms that sometimes echoed those of Tansley and other advocates of the Nature Conservancy in Britain. Dymond, for example, wrote in 1951 that "in the training of research workers it is essential that the attitude of scientific curiosity be retained and cultivated. If the inherent or aroused curiosity of the student is balked at every turn by the necessity of confining his interest to those activities which can be shown to have a good prospect of yielding facts of immediate practical application, the true spirit of science is very likely to be killed or badly dwarfed."[30]

This attitude reflected general aspects of Canadian science. Until the mid-1960s the National Research Council, the most significant research agency in Canada, one that was highly influential in setting science policy, had its agenda set largely by the scientists themselves. As in Britain and the United States, this stemmed from the status achieved by science through its contributions during World War II, and from nonscientists' willingness to defer to expertise in setting scientific priorities.[31]

While University of Toronto researchers stressed the long-term nature of Great Lakes fisheries research, and the need for its independence from immediate practical requirements, from the 1940s to the 1960s they nevertheless did make several specific contributions to management. For example, studies indicating the inefficacy of fish culture justified the closure of several hatcheries. Research at the South Bay station helped demonstrate the role of the lamprey in decimating lake trout and white sucker populations in Lake Huron.[32] Experimental fisheries and lake studies, as I have noted, also had implications for fisheries management. However, more direct links between research and management were established through expansion of provincial fisheries research, beyond its roots in University of Toronto studies. During the 1950s and 1960s the Department of Lands and Forests developed considerable research of specific relevance to management issues. Much research focused on species of greatest economic significance.

Studies by government biologists were often tied closely to local conditions. By the end of the 1950s there was a network of fishery biologists, distributed among twenty-two districts within Ontario, each responsible for advice and research on local fisheries problems. Much of the work of these fishery biologists consisted of responding to requests from the public for assessment of the fishing potential of particular lakes, or for "trouble shooting" of lakes where fisheries had failed. To gain greater knowledge about local conditions, an inventory program was begun, which surveyed fish populations and environmental conditions in about 1,000 lakes per year. Responding to a need for a standard method of assessing the productive capacity of lakes, Richard Ryder of the Department of Lands and Forests developed a "morphoedaphic index" relating lake morphometry and chemistry to potential fish production.[33]

Overall, therefore, the work of government fishery biologists was usually tied more closely to specific management issues than was that of university researchers. This reflected a basic difference in objectives. Seeking greater understanding of fundamental aspects of fish populations and their relationships with their environment, university scientists focused on detailed studies at a few sites. The government, however, needed to distribute research activities across the province, wherever specific management problems existed. As a result, the government was obliged to not rely on the university but to develop its own corps of biologists prepared to focus on specific, practical problems.

This difference in objectives may have been accentuated by the university scientists' insistence on their scientific autonomy. There is evidence that government administrators felt that the academic emphasis on fundamental research was not the most effective way to address fisheries issues. One illustration of this was a discussion at a 1957 meeting regarding the choice of a director of a new experimental station. According to the minutes of the meeting, "Dr. Fry maintained that the director should be a specialist with a Ph.D. degree in one of the fields of study planned for the Station. Mr. Johnston [head of the ODLF's Research Division] cautioned against the employment of a specialist who might be concerned only with the development of fundamental research in a specific field who would not necessarily provide for adequate consideration and/or attention to the direction of the other phases of the work planned for the station."[34]

One significant contribution that university scientists did make to management, however, was their view that fish stocks are controlled by environmental conditions, not by size of spawning stocks. This meant that fish stocks, even if depleted one year, could be relied on to restore themselves the following year.[35] And if it was almost impossible to permanently deplete fish stocks, then regulations controlling fishing could safely be relaxed, even discarded. Thus, researchers indirectly legitimized a policy of greater access to Great Lakes fish stocks and less regulation of fishing.

In the 1950s and 1960s the Ontario Department of Lands and Forests adopted such a policy, progressively loosening fishery regulations. From 1946 to 1960, Harkness, followed by C. H. D. Clarke, director of the Fish and Wildlife Branch during the 1960s, set as their objective the widest possible use of fish resources. The guiding principles of management became sustained yield, full use, multiple use, and public use.[36] Regulations hindering full use of the fish resource were considered to be as great a danger as overfishing. A. G. Huntsman, another University of Toronto biologist, summarized the danger posed by regulations imposed without adequate proof of their need: "It is well established that the stocks of fish in both ocean and inland waters fluctuate in amount, sometimes very greatly, over shorter or longer periods. Time after time, these changes in abundance have been wrongly interpreted as caused by man's action. If we are to avoid applying a wrong remedy, these fluctuations will have to be clearly recognized. If the take of fish decreases from such cause from year to year, if such decrease is attributed to over-fishing and if consequently the take is restricted by regulation, we are needlessly making a bad situation worse."[37] The burden of proof, therefore, lay on those who favored regulations: in the absence of strong evidence that regulations were required, they would not be imposed, or they would be relaxed or removed.

Not all scientists, it must be noted, believed that overfishing was unlikely. In 1962, Henry Regier, at the time a fisheries biologist for the Department of Lands and Forests, had suggested that overfishing was the likely cause of fluctuations or depletion of Lake Erie walleye. A few years later he and a colleague found evidence that declining whitefish populations in Georgian Bay, and greater instability, were the result of an intensive fishery. In 1963, Jack Christie, a biologist at the department's Lake Ontario fisheries station, suggested that overfishing was damaging that lake's whitefish stocks. He repeated this warning during the 1960s and early 1970s, and he argued against liberalized regulations on the grounds that overfishing could destabilize fish populations by upsetting competitive and prey-predator interactions between them.[38] Ineffective fisheries regulations, he noted in 1968, limited biologists to describing the progressive elimination of fish stocks, without knowing why these have occurred.[39] However, throughout the 1960s the dominant assumption remained that overfishing was usually not of immediate concern.

Besides liberalizing regulations, the Department of Lands and Forests, as well as the federal government, acted on this assumption by subsidizing fisheries expansion, introducing new gear and other technology, and exploring for unexploited stocks. When, for example, Lake Erie whitefish, walleye, and other valuable species declined in abundance (resulting in much economic hardship in fishing communities), and smelt populations increased, the province and the federal government subsidized fishermen's purchase of smelt fishing gear. Science was invoked to justify these policies: research, Clarke and other officials noted, had found that declines in fish stock and shifts in dominant species were caused by changes in the physical environment, not overfishing. Those unconvinced were perhaps reluctant, they suggested, to accept scientific results; as the department's magazine editorialized, "While it has been found, time after time, that competition and over-fishing are more myth than reality, they are concepts difficult to dispel even in the light of scientific evidence."[40]

This view was one aspect of a more general attitude toward the factors that control animal populations, and the practical implications of this. Clarke, for example, was already receptive to such views because of his own background in wildlife management. He believed that destruction of habitat had a far greater impact on wildlife than hunting, that demands for regulation reflected inappropriate application of the lesson to be drawn from the passenger pigeon and the bison, and that the "new science of animal numbers has taught us that an area can only support a certain number of animals and that if the surplus is not harvested it is wasted."[41] More generally, such policies reflected the view, widely held until about the mid-1960s, that the appropriate role for Canadian governments in natural resources management was to assist in their rapid exploitation.[42] In providing justification for a policy of loosening fishery regulations in Ontario, scientists were therefore consistent with this perspective.

DECLINE OF A RESOURCE

By the late 1960s, however, provincial fisheries policies were coming under increasing pressure, as it became apparent that they could not cope with the intense and diverse stresses imposed on the Great Lakes fisheries. Fish stocks had exhibited alarming declines. The Canadian Lake Erie walleye (often called pickerel in Ontario) fishery, after peaking at 4.2 million kg in 1956, declined to insignificance within five years. Whitefish in Lake Ontario exhibited a similar crash in the 1960s, just as the lake trout had twenty-five years before, and the Atlantic salmon had a half century before that. By 1970 the Ontario commercial fishing industry was relying almost entirely on species of lower value, such as yellow perch and smelt. As a result, while the quantity of fish landings remained at historically high levels, their value declined, leading to much hardship in fishing communities. Other activities dependent on high-quality fish, such as professional guiding of angling parties, virtually disappeared; and the loss of species exacerbated conflicts between commercial and sport fishers. While in the early 1970s the take of yellow perch from Lake Erie was widely considered unsustainable, federally subsidized investment in ships and wharves continued even though the resource could not be increased, thereby reducing real productivity.[43] Overall, the commercial fishery was far from economically productive or sustainable.

The decline of the fisheries had several possible causes. One was over-fishing; in the 1960s some researchers, as I have noted, were identifying it as a factor in the depletion of fish stocks. Another was domestic and industrial pollution, as reflected in concern about phosphorus inputs and eutrophication of Lake Erie, and to a lesser extent, Lake Ontario.[44] Toxic substances were also exerting some effect; in 1970, for example, the commercial fishery in Lake St. Clair was closed because of mercury contamination. The intentional or accidental introduction of new species, including the lamprey, alewife, and carp, had also led, some believed, to changes in fish populations.

But faced with these multiple possibilities, there was no consensus on the causes of the declines. Provincial fisheries managers, who considered the environment the dominant factor in determining fish populations, most often blamed degradation of the environment for the precipitous declines in stocks. Some considered exotic species that preyed on native stocks (the lamprey) or outcompeted with them for food (the alewife, for example), to be significant. A few researchers, as I have noted, identified overfishing as important.

With a declining fishery, increasing conflicts over remaining stocks, an array of possible causes of the decline, and the fact that some causes, such as pollution, were not even within the jurisdiction of the Department of Lands and Forests, it became apparent that existing approaches to fisheries management were not effective. Management goals had become limited to slowing the retreat, ensuring maximum use of those species that remained, or, as Christie described it, "making best use of each stage of the deterioration of the fish stocks." Clarke expressed his own doubts, describing a "very uncertain future" in which no species, whatever its numbers, could be considered secure.[45] Standard approaches to fisheries research could not provide a way out. Studies of individual species were clearly inadequate for understanding of the complexities of the fisheries. Most important, they could not assess the relative significance of the various factors (exploitation, pollution, exotic species) damaging the fisheries. Lack of consensus on this point hindered action on all of them. Overall, the Great Lakes, as the Canadian biologist Douglas Pimlott noted in 1971, were an "epic case of man's mismanagement of natural environments . . . [with the] fishing industry in a depressed and chaotic state. . . . Canada may well be committed to an ill-advised pattern of scientific activity."[46] And Peter Larkin, a Canadian fisheries ecologist, noted the irony of the "story of the fisheries on the Great Lakes, a bewildering tangle of regional regulations, introductions of exotic species, and environmental effects, which is of no value as an example of how to manage, even though the Great Lakes have long been adjacent to a strong concentration of fisheries biologists." And so by the late 1960s there had emerged a sense of urgency to develop a new approach to fishery management and research.[47]

AN ENVIRONMENT FOR CHANGE

Even as the Great Lakes environment was changing, so were the links between the provincial government and University of Toronto fisheries researchers. The government's own expertise, built up since the early 1950s and composed in large part of scientists and managers trained at the university, as well as graduates of a fish and wildlife management program at the University of Guelph near Toronto, made it less dependent on the university for scientific guidance and legitimation. By the 1960s the academic scientists' role in setting provincial research priorities had been assumed, to a large extent, by government scientists, fisheries managers at ministry headquarters and in field offices, and public demands for study of specific problems.[48] This shift was also evident in a restriction of university scientists' control over funds provided by the government for research: in 1968 it was decided to end institutional support for outside research contractors, including universities; support would instead be provided for specific projects only.[49] On the university side, the architects of the original government-university partnership were leaving the scene. Dymond, Fry, and Langford had all retired by 1972. They were replaced by a new generation of aquatic ecologists hired during the university's expansion of the 1960s. This expansion included a new building for the department of zoology, providing greatly enlarged facilities for ecological research.

These ecologists began work in several areas of research new to the University of Toronto. The work of two of them, Frank Rigler and Harold Harvey, exemplified this development. Rigler, a chemical limnologist, studied the dynamics of phosphorus in lakes. By the early 1970s he had developed a highly empirical perspective on this topic and was seeking to relate phosphorus concentrations to the growth of algae in lakes. He and a student, Peter Dillon, demonstrated the practical implications of this work by developing a model to predict lake eutrophication and deterioration in water quality on the basis of nutrient inputs. After graduation, Dillon continued this work at the Ontario Ministry of the Environment.[50] By 1975 Rigler had left Toronto to join McGill University in Montreal.

Like Rigler, Harold Harvey had arrived at the University of Toronto in the mid-1960s. He initially pursued studies in limnology, including the relation between physical and chemical characteristics of lakes and fish populations, as well as other aspects of the physiological ecology of fish.[51] By the end of the decade, however, he had begun a new area of research. It began serendipitously in 1967, when he discovered that several lakes near Sudbury had both impoverished fish populations, and were more acidic than expected. In 1969 he and a student, Richard Beamish, began recording the acidity of 150 lakes in the region, as well as the acidity of local precipitation and monthly wind directions, to evaluate the possibility that the cause of this acidity was sulfur dioxide released by metal smelters in Sudbury. At the time, these smelters were one of the largest sources of sulfur dioxide in the world, with 2.6 million tons released in 1969 alone. Fish species in several lakes were also surveyed each year, identifying a pattern of disappearance of their populations as their lakes became acidified.[52] Acid precipitation and its impacts on lake chemistry and fish populations became for Harvey a research focus for most of the subsequent two decades. Beginning in the late 1960s he also publicized the issue of acid precipitation, through the media and talks before interested groups. His results, and those of other researchers in Canada and elsewhere, would lead, in the 1980s, to effective controls on Canadian emissions of sulfur dioxide.

Rigler's work on the impacts of nutrient inputs into lakes, and Harvey's extensive studies of acid rain, were two reflections of an interest in the practical implications of ecology and of a sense of responsibility to contribute, as university scientists, to public debates, that was strongly felt within the department of zoology. One of Canada's most prominent environmental groups, Pollution Probe, originated in the department in 1970. Donald Chant, the department chair, was also an environmentalist who cowrote, with his colleague Ralph Brinkhurst, a survey of environmental problems for nonscientists.[53] Henry Regier, who joined the department in 1966, was also a strong environmentalist who was motivated by an "ecological conscience" that accepts that "natural communities have rights."[54]

Collaborative projects involving several scientists also became a more significant focus of research activity at the University of Toronto. In Canada, as in the United States, Great Britain, and many other nations, the International Biological Program provided opportunities for such projects. Under the aegis of the Canadian Committee for the International Biological Programme, ecological studies of arctic, woodlands, grasslands, and agricultural terrestrial habitats, and of marine and freshwater (temperate and arctic) aquatic habitats, were conducted between 1968 and 1973.[55] University of Toronto ecologists led one study, of Char Lake, on Cornwallis Island in the high arctic. At the same time, a parallel opportunity for collaborative research was provided by the Fisheries Research Board of Canada (FRBC), in the form of substantial grants to three universities: the University of British Columbia, the University of Manitoba, and the University of Toronto, for research on various aspects of the aquatic environment.

As Rigler, its director, explained, the Char Lake project had both practical and theoretical justifications. It was a case study of the sensitivity of an arctic lake to the environmental impacts of human activity in the north. The simplicity of its biological communities would also make generalizations concerning productivity and energy flow more readily obtainable than in more complex ecosystems.[56] The focus was on the productivity of all components of the aquatic ecosystem and on identifying factors responsible for the low productivity characteristic of this and other arctic lakes. A major difference between this and IBP projects in the United States was that less emphasis was paid to integration of results. Published reports focused on particular ecosystem components—fish or invertebrates, for example—not on characteristics of the entire ecosystem. Ecosystem modeling was not a significant element. Some attention was devoted in the Canadian IBP to computer modeling, including a workshop for project directors; some students at Char Lake also attempted to develop a whole lake ecosystem model.[57] However, this attempt was unsuccessful, and models generally were much less significant than at Oak Ridge. The conditions at Oak Ridge that encouraged a heavy emphasis on modeling—prior interest, experience, and facilities—were all lacking in the zoology department at the University of Toronto. Instead, the project provided researchers with an opportunity to extend their individual research interests within a new environment.

In response to a consolidated proposal submitted in 1968 by several Toronto ecologists, the FRBC provided support until 1973, when it terminated its practice of substantial funding for university researchers. In their proposal, ecologists had stressed the practical implications of their research, particularly in terms of fishery biology, thereby asserting its relevance to the board's mandate. In practice, this funding enabled university ecologists to continue their own research into several aspects of the aquatic environment, including, for example, Rigler's studies of phosphorus dynamics and Harvey's research into acid precipitation.

The Char Lake project and the support provided by the FRBC were followed by other research initiatives within the zoology department. In 1975 several aquatic ecologists and limnologists initiated the Lake Ecosystem Working Group. It involved joint research proposals, collaborative studies, seminar series, a course for senior students, and other activities intended to provide a synthetic perspective on aquatic ecosystems. Numerous smaller projects were also conducted within the department, including a study of a Lake Erie fish population funded by the Great Lakes Fishery Commission, and another of the impact on Lake Huron fish populations of heated effluent from a nuclear plant, supported by Ontario Hydro (the provincial electrical utility).

The evolution in the relationship between ecology and environmental concerns that occurred at the University of Toronto was similar in several respects to contemporaneous changes at the Oak Ridge National Laboratory. In both cases there were new funding opportunities for ecology: the IBP in both countries and the Fisheries Research Board in Canada. Ecologists also developed a greater commitment to theoretical issues of interest to academic ecologists elsewhere and distanced themselves from the practical issues (radiation at Oak Ridge, fisheries at Toronto) distinctive to each site, exploring instead other environmental issues. One difference, however, was the more substantial growth of environmental research at Oak Ridge than at the University of Toronto. Whereas American federal agencies turned to Oak Ridge for much of their environmental research needs, the Canadian federal government chose instead to establish a new aquatic research facility (the Canada Centre for Inland Waters) on Lake Ontario, 60 kilometers from the university.[58] In the early 1970s it also began directing much of its environmental research funding to consulting firms in an effort to develop a Canadian environmental consulting industry. But the fact that University of Toronto ecologists did conduct some specific projects for other agencies, such as Ontario Hydro, contrasted with the decision of Hubbard Brook ecologists not to do so. Financial stringency appears to have been a major reason: in the absence of steady support from a national research agency after the end of the IBP and Fisheries Research Board projects, ecologists in Toronto had to actively pursue funding from a variety of sources to maintain their research programs and support graduate students.

For its part, the Ontario Department of Lands and Forests (after 1972, the

Ministry of Natural Resources) became less dependent on the university for scientific advice, research direction, and legitimation of its policies. One consequence of these developments was seen in 1972, when the university and the provincial government canceled their cooperative agreements for the Maple and Lake Opeongo laboratories. These cancellations had an immediate, practical justification: the province needed more space at both laboratories, and the university had just gained a large new zoology building and no longer needed the space it was using. But they also reflected shifts in practical and theoretical commitments, and looser ties between the university and the provincial government. These ties were not severed: university and government scientists continued to cooperate by sharing boats, nets, and ideas. Such arrangements reflected the opportunities for pragmatic sharing of resources that occurred with separate research entities pursuing activities within the same region. They differed considerably from the formal ties of advice, policymaking, and legitimation that had characterized relations between the university and the government.

These looser ties, and the disengagement of university research from its role in justifying fisheries management policies, also provided an opportunity for a new relationship between the university and the provincial government. The opportunity was taken up by Henry Regier (see fig. 9).

After receiving his Ph.D. from Cornell University in 1962, Regier had been hired by the Department of Lands and Forests for studies of the Great Lakes fisheries; he continued this work after being hired by the university. His initial studies, of the dynamics of individual species populations, were compatible with the dominant perspective at the university; in fact, as he recalled, he was hired in part because of this work.[59] By 1967, however, he had begun to view such an approach as incomplete. Studying the Lake Erie walleye fishery, he realized that its collapse in the 1950s had several causes, including environmental degradation, overfishing, and interactions between fish species.[60] To Regier, this implied the need for an approach to fish populations that was better able to deal with the complexity of the environment.

While completing his work on the Lake Erie walleye, Regier began planning a large comparative study of fish communities in lakes. This culminated in the Salmonid Communities in Oligotrophic Lakes Symposium (SCOL symposium) in 1971.[61] Researchers from Ontario, the United States, and Europe brought together studies of fish communities experiencing one or more stresses (including exploitation, eutrophication, and exotic species). By comparing fish communities subject to no major stress with those subject to one or more, they inferred general patterns describing the response of fish communities to these stresses.

Fig. 9. Henry Regier (courtesy of Henry Regier)

A major conclusion of the SCOL symposium was that fish communities exposed to different stresses tend to behave in similar ways. For example, the fish community of a pristine lake will usually be dominated by a few larger, longer-living, predominantly bottom-dwelling fish species, the abundance of which remains roughly stable from year to year. But when the lake is subjected to a stress, such as exploitation or excessive nutrients, the fish community will tend to be transformed into one dominated by less stable populations of smaller, pelagic fish species with shorter life spans.

These conclusions, as developed by Regier and colleagues, drew on the insights of ecologists elsewhere. Eugene Odum, Ramon Margalef, and others, Regier noted, had suggested that such patterns of response to stress could be observed in communities and ecosystems.[62] A chief contribution of the SCOL symposium was to demonstrate that these patterns could also be used to understand changes in fish communities. Fisheries research could therefore, Regier contended, both learn from general ecological theory and enhance its own contribution to that theory.

Some provincial fisheries managers and researchers were predisposed to these ideas, and they contributed to their formulation. For example, Jack Christie contributed insights on the competitive interactions between fish species and their role in maintaining the stability of the fish communities of that lake.[63] His own research supported a chief implication of the SCOL symposium—that studies of the dynamics of individual species should be supplemented by studies of the effects of stresses, including changes in interspecific relations, on entire fish communities and ecosystems. Christie's receptiveness to these ideas was perhaps also enhanced by his frustration with existing management approaches.

Ken Loftus, a fisheries manager since the late 1940s, also found these ideas of great interest, and he coedited the symposium with Regier. Loftus became director of the Sport Fishing Branch in 1973, and in 1977 he was put in charge of provincial fisheries management, so he was in a strong position to develop the practical implications of this research. The most important implication was that the objective of fisheries management should not be maximum yield of a few species, but a balanced fish community with a diversity of species able to maintain stability from year to year. This implied a different view of changes in fish populations: where once, for example, a large increase in a fish stock was a bonanza, it now indicated instead that something was wrong, that the fish community was not stable. Managers also had to be aware of the impacts of fishing and other stresses not just on species of immediate interest to humans, but on all species, because all may contribute to the stability of the community.

These developments in Ontario mirrored ideas evolving among fisheries scientists elsewhere. In Michigan, Stan Smith had demonstrated the significance of interactions among fish species to the stability of fish yields.[64] More generally, there was a growing sense within the international fisheries science community that current strategies of fisheries management were inadequate. Maximizing yield of a single species often led, it was realized, to greater instability and depletion of fish stocks. There was also a growing awareness of the need for better ways of dealing with uncertainty and the complexities of aquatic ecosystems, and to better integrate the technical aspects of fisheries management with socioeconomic concerns, such as the allocation of fish to different sectors of society or the impact of economic factors on the behavior of fishermen.[65]

Two years after the SCOL symposium, in February 1974, scientists and fisheries managers from the Ontario Ministry of Natural Resources and the Fisheries and Marine Service of the Federal Department of the Environment formed a working group to develop a new strategy for Ontario fisheries. They began with the premise that the deterioration of fish stocks over the preceding decades amply proved the inadequacy of existing management practices. The problem, they believed, was reliance on fragmentary perspectives. Scientists had provided narrow and mutually incompatible advice: "Our inability to protect and conserve the native fish communities of the Great Lakes can be traced to the fact that too many experts had simple but different conceptions of the problem of collapsing stocks. We failed to recognize that experts from different disciplines had conceptions of the problem that were partially correct. A broader view of the problem might have allowed effective action on several fronts."[66] The need, they asserted, was for a broader, more comprehensive perspective on fish management as a basis not for asserting the demands of different competing interests, but for addressing the priorities of society as a whole.

This perspective would acknowledge the complex ecology of fish stocks—the behavior of fish communities, the impacts of fishing, pollution, exotic species, and habitat destruction—as well as the notion that fisheries management was not only an ecological but also an economic issue. Management had often relied on simplistic assumptions about the economic behavior of fishermen, concerning, for example, their self-regulation.[67] It would also consider all the uses and values attached by society to fish stocks. As a source of recreation, and as an indicator of the health of the aquatic environment, fish were no longer of concern only to fishermen. Fisheries agencies now had a broader responsibility to society. Fulfilling this responsibility demanded a step off the treadmill of reactive decisionmaking and responses to crises, and a move toward proactive planning, articulation of goals, and devising of management strategies that consider "all major benefits, all major users, the limits of supply and uses of waters antagonistic and complementary to fisheries."[68]

An essential element of this new strategy would be a change in the role of science in fisheries management. Until the late 1960s, Ontario fisheries scientists generally defined their task in terms of the resource use of highest priority: production of fish flesh. Their role was to ensure maximum production by predicting, as precisely as possible, the size of stocks of species of greatest interest to fishermen. Other aspects of Great Lakes fish and their environment—fish community dynamics, impacts of other human activities—were abstractions into which the study and application of population dynamics would not stray. The problem was that this approach provided no way to assess the relative impact of different stresses—not only fishing, but, for example, pollution, exotic species, or habitat loss—on fish communities. As a result, in the face of conflicting scientific advice concerning the causes of collapsing fish stocks, all the parties to an allocation debate often invoked conservation as the basis for their point of view, and the ultimate effect was nearly always further depreciation of the resource.[69]

In contrast, fisheries ecology, as represented in the SCOL symposium and elsewhere, defined its role in the context of a diversity of uses of the environment. It would trade the potential (never realized, in practice) to predict future fish stocks, for the capacity to understand and compare the impacts of several simultaneous uses, establishing that certain such uses—waste disposal, or transformation of habitat, for example—were most damaging and should be restricted in favor of more benign human activities. The lesson drawn from this was that science should be used not in asserting the claims of particular users but to identify the conditions necessary for conservation of the resource. Conservation was a matter of general concern to society, because if there were no resources, there could be no users. Only once conservation had been assured could there then be negotiation among users concerning its allocation. Thus, the task of fisheries management itself was shifted from ensuring maximum yield by particular fish stocks to the more complex task of coordinating and allocating resources to a variety of users, guided by an understanding of the impact of these uses on the entire fish community and its environment, as provided by ecological research.

Between 1974 and 1976 these ideas were put into practice, as provincial and federal policymakers and scientists pursued a strategic approach to fisheries management. They began by defining the values and benefits associated with the fisheries, using these as a basis for a comprehensive set of goals for federal and provincial agencies:

- Protection of the human environment.
- Conservation and enhancement of fish communities and habitat.
- Human health and healthy fish communities.
- Optimization of recreation, income, and employment.
- Harmonious use of fishery resources and habitat.[70]

Critical to achievement of these goals, the working group emphasized, was protection of the fish resource itself: with no fish, there could be no benefits. Accordingly, protection of the aquatic environment and its fish communities would come first, and allocation of fish to fulfill other goals had to be secondary.[71]

The working group's next steps were to formulate objectives, defining more specifically these goals, and then to work out strategies to achieve them. These strategies, eventually formulated as the Strategic Plan for Ontario Fisheries (SPOF), replaced the previous emphasis on open access to fish stocks, with new strategies based on limited access, explicit allocation of fish stocks, preference for sensitive uses (that is, that require a healthy environment) over degradative uses, public education and participation in management, and, above all, protection of fish communities. Overall, the strategy would foster maintenance of the generally intact fish communities of the north and reha- bilitation of the degraded fish communities in southern Ontario, including Lake Erie and Lake Ontario.[72] SPOF was completed in 1976, won approval from the Ontario cabinet two years later, and was implemented over the fol- lowing decade by the Ministry of Natural Resources. During that time the fish stocks in the Great Lakes made some recovery, including walleye and yellow perch in Lake Erie, yellow perch and whitefish in Lake Huron, and lake trout and whitefish in Lake Superior.

Recent developments in fisheries science helped shape and justify SPOF. The working group repeatedly noted the consistency of its conclusions with those of the SCOL symposium: the value of the diversity and stability of natu- ral fish communities, their status as indicators of environmental health, the need for an integrated perspective encompassing all stresses on fish commu- nities and their environment, and the possibility that if these stresses could be reduced, rehabilitation of these communities would be possible. SCOL was described as a "turning point for management science": besides these lessons, it showed that lakes could be studied in sets, not as individual, unique entities. Thus, it provided a basis for experimental management, in which lessons learned on one lake could be generalized to other, similar lakes.[73]

Overall, therefore, an ecosystem perspective, its relevance to the fisheries demonstrated through the SCOL symposium and related activities, provided an alternative basis for management. It did so by demonstrating that a variety of stresses—fishing, eutrophication, exotic species—affected fish stocks; that these stresses, together or separately, had broadly similar effects; and that fisheries management, to be effective, could not focus on one stress (or one species) to the exclusion of all others. Instead, all stresses must be moderated sufficiently to ensure survival of fish communities composed of healthy, stable stocks of desirable species. Only recognition of the impact of all stresses on fish communities could provide a way out of the unproductive debate over which was most important. The role of SCOL in shaping the Strategic Plan for Ontario Fisheries is not surprising, as several individuals were important in both efforts. Ken Loftus, coeditor with Regier of the SCOL symposium, was cochairman of the SPOF steering committee, and Regier served as an adviser throughout the process. Their roles demonstrated the significance of direct interaction between scientists and policymakers within government.

But the design of the strategic plan, and its eventual adoption by govern- ment, was a result not only of scientific arguments and close links between managers and scientists. Another factor was the pressure for changes in fisheries management practices, generated by the decline in the fisheries and by increasing public concern over the environment. An approach embedding fisheries management within a larger environmental protection effort helped ensure its favorable reception by a government eager to demonstrate some response to this public concern. Interest groups more directly concerned with the fisheries were also receptive. Commercial fishermen had become more willing to acquiesce to new policies simply because their most profitable fisheries had collapsed and they had less to lose. Sportfishers were also receptive because SPOF promised greater emphasis on the larger fish species, like salmon, that they preferred.

SPOF was also, in part, the product of an interest in comprehensive planning. In the early 1970s the federal government launched numerous studies of the potential benefits of water resources in river basins and other areas, the means by which these benefits could be achieved, and the objectives that might thus be served.[74] Like SPOF, these were comprehensive exercises, aimed at developing optimal strategies for balancing the uses of water bodies. Other initiatives, including a review of national fisheries research and management activities, also reflected federal interest in strategic planning.[75] The Ontario government demonstrated a parallel interest in comprehensive planning and decisionmaking, in part by reorganizing in 1972 the Department of Lands and Forests, which it renamed the Ministry of Natural Resources.[76] SPOF, therefore, also reflected an interest in comprehensive management strategies in both the federal and provincial governments. Such strategies would be based on comprehensive scientific understanding of the environment and would seek to allocate various uses of environmental resources in order to reduce conflict between resource users.

This new approach may be compared with contemporaneous innovations in California's coastal fishery, as described by Arthur F. McEvoy.[77] These cases were similar in several respects. In both, declining fish stocks, conflicts over their use, and more general shifts in perceptions of the environment and its value combined to foster new management strategies. These strategies were responsive both to the status of fish stocks and to the requirements of other species: whether they were fish that help maintain healthy, stable fish communities in the Great Lakes or pelicans and otters fulfilling a similar role off California. And in the Pacific, as in the Great Lakes, these strategies demanded revision of the role of science: from estimating maximum yield to providing the basis for maintaining or restoring diverse aquatic ecosystems. They also implied recognition of regulation as not simply a technical process but a political and economic one requiring adherence to democratic norms of broad participation and negotiation if its conclusions were to be accepted by resource users.

Ontario fisheries scientists long viewed fisheries management as primarily a technical problem: accurate prediction of fish stocks, once achieved, would permit effective management and maximum production. But in the Great Lakes region, where natural ecosystems coexist with a large human population and intense industrial activity, fisheries management had become only one aspect of the larger problem of resolving conflicts between different uses of the environment. This problem was as much social and political as technical. Its solution, in the view of the scientists and policymakers developing the strategic plan, demanded a comprehensive perspective on uses of the environment, as provided by an ecosystem-based approach to fisheries science.

As we have seen, while these changes in fisheries management attitudes were occurring, ecological research at the University of Toronto was also being transformed, as was the relation between university research and the provincial government. Close ties between natural resources management and university scientists have been an important feature of the recent history of Canadian natural science; this was clearly in evidence in the role played by University of Toronto scientists in guiding and justifying fisheries management policies in Ontario. Changing social priorities and environmental conditions eventually made a new management approach necessary. The role that university scientists played in formulating this approach reflected the erosion of the relationship between university and government, their weak allegiance to the scientific perspective that had been the basis for this relationship and their development of a new ecological perspective better able, within the Canadian political context of the 1970s, to address these new priorities and conditions.

8

Comparing Ecologists and Their Institutions

Between 1945 and 1980 the world of ecologists was transformed. Many more research opportunities emerged as ecology programs in universities, government agencies, and elsewhere proliferated in the postwar remaking of civil science—its funding, organization, and social roles—in Great Britain, the United States, and Canada. The ideas and practice of ecology were also transformed as research perspectives coalesced around influential leaders, merged, and reformed. We have encountered several of these, such as production ecology, radioecology, and ecosystem ecology. One consequence was much greater ambiguity concerning the identity, even the existence, of central ideas and methods that all ecologists could agree upon. To a large extent ecology was indeed "in the eye of the beholder," as Robert McIntosh has remarked.[1]

In preceding chapters I have described how ecologists at several institutions contributed to this development of contemporary ecology. Before discussing these episodes, and what they can tell us about the place of ecology in environmental politics, I will review our current understanding of this topic.

PERSPECTIVES ON ECOLOGY AND SOCIETY

A persistent theme in ecology, and in studies of its history, has been the reciprocal ties of inspiration and legitimation of values between ecologists and their society. In *Nature's Economy,* Donald Worster argued that the utilitarian, managerial values dominant in society are echoed in contemporary ecology. Important episodes in its history—the progressive conservation movement and its ideology of maximum resource productivity; wildlife management, which in the 1920s and 1930s absorbed this ideology; studies of production and consumption in ecological communities, based on a foundation laid by Charles Elton; and, most recently, studies of energy flows in ecosystems—all have been intimations of the New Ecology, which "sees nature through a different set of spectacles: the forms, processes, and values of the modern economic order as shaped by technology."[2] Other historians have also seen reflected in ecology certain social values and attitudes. Gregg Mitman, in his analysis of the work of Warder Clyde Allee and his colleagues at the University of Chicago, linked their belief in a peaceful society to their scientific understanding of cooperation within ecological communities. Peter Taylor found in Howard T. Odum's energy flow studies the technocratic optimism prevalent in some American political and scientific circles between the 1940s and the 1960s. Finally, Chunglin Kwa has noted the consistency of the cybernetic metaphor used by ecologists promoting the International Biological Program, with the technocratic assumptions of American policymakers.[3]

These interpretations of the history of ecology in terms of basic attitudes and values of society have their counterpart in our understanding of the social role of ecology: as an expression and scientific justification of attitudes and conduct toward nature. For example, Worster structured his account of the history of ecology in terms of the contradictory attitudes expressed by ecology at different times: the "arcadian" perspective, implying respect for the intrinsic value of nature, and the "imperialist" view of nature as a storehouse of economic resources. Leo Marx has also defined the role of science in terms of its support for society's basic attitudes toward nature, though he identifies three such attitudes: progressivism, primitivism, and pastoralism.[4]

Extending this interpretation, it has often been argued that ecologists' social role has been expressed through their efforts to instruct society in the implications of their work. Indeed, the emergence of the environment as a public concern has been depicted as a product of ecologists' success in communicating their message. As Joseph Petulla has argued, "ecological science . . . has provided an intellectual basis for and a predominant role in the leadership of the contemporary environmental movement."[5] And, as Worster stresses, the mere existence of environmental concerns places ecologists in a central role:

> In recent years it has become impossible to talk about man's relation to nature without referring to "ecology." This peculiar field of study has been suddenly called on, in a manner unusual even in our science-impressed age, to play a central intellectual role. Such leading scientists in this area as Rachel Carson, Barry Commoner, Eugene Odum, Paul Ehrlich, and others have become our new delphic voices, writing bestsellers, appearing in the media, shaping government policies, even serving as moral touchstones. So influential has their branch of science become that our time might well be called the "Age of Ecology."[6]

Similarly, Thomas Dunlap has identified a central role for ecology in shifting society's attitudes toward nature:

> In the nineteenth century evolutionary biology told us of our origins and showed us how we were connected to other forms of life. It challenged, then displaced, religion as the source of knowledge about the world. In the twentieth century, ecology provided a complementary vision. It explained how organisms interacted, how the "economy of nature" really worked. Science has been the guide—at least rhetorically—for wildlife and nature policies and for the picture of nature we receive in books, movies, and television shows. It justifies Romantic identification with nature animal rights. It is, in short, the authority to which we turn for new ideas and ratification of old ones.[7]

Implicit in this interpretation is the view that environmental concerns reflect the dissemination of certain values into society, those who first understood these values teaching the rest of society. Ecologists, it has been stressed, have played a crucial role in this public education. Through their knowledge of the stability, balance, and diversity of nature, ecologists have shown humanity how to minimize its interference in nature and how to reshape nature to serve society.

This interpretation has influenced historians' analyses of the work of particular ecologists. Consider, for example, the American ecologists Victor Shelford and Howard T. Odum. Shelford believed strongly in ecologists' responsibility to act on the practical implications of their research. His concern about "people's shortsightedness and unconcern for the damage being done to natural resources" led him to study the effects of industrial wastes on fish, and to advocate preservation of natural areas and inclusion of ecological principles into park and forest management practices. In contrast, Odum's goal was ecological engineering. By analyzing ecosystems in terms of energy circuits he thought it would be possible to understand the cybernetic mechanisms that ensure their stability and optimum performance. This understanding would be the basis for their efficient management.[8]

Obviously, Shelford and Odum had very different prescriptions: the preservation of nature, or its efficient management. But historians have interpreted their social roles in similar ways: in both cases, it is assumed that their roles are derived from the implications of their work for society, and their willingness and ability to elucidate these implications. Accordingly, to understand these roles it is necessary to focus on the underlying values and attitudes expressed in the work of these ecologists, and not, for example, on their institutions, except to the extent that these provide or deny research opportunities.

This perspective is also evident in analyses of the evolving political significance of ecologists: as their ideas change, so does their role in society. According to Hagen, for example, ecologists' loss of confidence in their capacity to solve environmental problems reflects pessimism concerning the prospects of a unified ecology, and skepticism of the intellectual merits of ecosystem ecology. Similarly, Worster has argued that ecologists' diminished willingness and ability to teach society stems from changing ideas within their discipline. With Clementsian succession theory and the ecosystem concept in decline, stability and order are no longer evident in nature; instead, ecologists see disturbance, instability, and chaos. The lesson of the older ecology was evident: do not disturb the balance of nature. It is more difficult for ecologists to draw as clear a lesson from the new ecology.[9]

If we extend this interpretation to the present work, we can see how ecological research at each of the sites I have examined expressed certain social values and attitudes. For example, fisheries research in Ontario exemplified the resourcist values of maximum productivity for human benefit, which Worster and others have described as a defining feature of both the New Ecology and industrial society. Some ecosystem simulation models developed at Oak Ridge exhibited the technocratic tendencies that, according to Taylor and Kwa, exist in both ecosystem ecology and in the ideology of expertise that fostered it. The view of conservancy ecologists, that nature reserves could be best maintained through active intervention, and their belief that many supposedly natural areas had in fact been modified by humans, reflected a longstanding perception that the British people and their landscape have a common history and identity.[10] Such interpretations remind us that the representation of nature provided by science reflects not only "objective" data but the context of scientific activity, including social values.

However, a focus on the underlying implications of ecology hinders the consideration of ecologists on their own terms, in their own times, situating them instead within the evolution of basic value systems. For example, eco-centrism has been described as originating in the writings of Henry David Thoreau, John Muir, and other romantic transcendentalists, and technocentrism as originating in the American progressive conservation movement. The assumption is that once these essential values were expressed they remained essentially unchanged. They are, in effect, ahistorical values to which subsequent ecologists have declared their allegiance or their opposition. But it is necessary to see ecologists not simply as players in a larger historical drama, acting within a panorama of evolving ideas and values, but as individuals guided by their own background, observations, ambitions, and the constraints and opportunities defined by their scientific disciplines.

To understand ecologists' place in society it is also necessary to consider not only the values expressed in their work but also those implied by the contexts in which they work: the emergence of the environment as a public concern, the evolution of specific environmental priorities, and the perceived relevance of ecological expertise to these priorities. Also important is the general political context, including prevailing views concerning the conduct and roles of government, the place of scientific expertise in society, and the organization and direction of scientific activity. Finally, it is necessary to examine the role of institutions in mediating the interaction between ecologists' own priorities—both scientific and personal—and the priorities of their society. Examination of its political and institutional contexts can not only help us identify the values implicit in ecology, it can also help us understand how, specifically, these values are communicated between ecologists and their society. Only then can ecology be realistically interpreted as, like other sciences, the product of an integration of context and cognition.[11]

THE GROWTH OF ECOLOGY: A FIRST LOOK

Let us begin by considering some general implications of environmental politics in the context of the growth of ecology since the 1940s. Most obviously, this growth can be understood in terms of numbers: the expanding corps of ecologists at each institution mirrored the expansion of ecology in each country.[12] A less obvious dimension of growth has been that of a national (and international) discipline within which ecologists achieve success by developing concepts and methods of interest to their colleagues everywhere, not just those within their own region, or who study similar species or natural communities. Instances of this are readily apparent. While the conservancy, recognizing the specific and unique problems of each area of Britain, developed a regional organization for conservation research, it also provided opportunities for ecologists like Ovington, who ranged over England and across the Atlantic, developing ideas about forest ecology of more general significance. At Oak Ridge, ecologists gradually expanded their research from a contaminated forest and lake bed to the southern deciduous forest to watersheds and pastures representative of environments throughout the country. They sought to develop methods of interest not just to ecologists studying radioactivity but to all ecologists. Hubbard Brook ecologists, in part by comparing their results with studies elsewhere, aimed to understand not just their own forest but northern hardwood forests generally. University of Toronto fisheries ecologists concentrated on a few research sites, seeking results of general application, while the provincial government, in contrast, developed a capability to assess conditions in specific lakes and streams. In each case ecologists directed their attention to sites intended to be widely representative, deemphasizing the unique features of any one site.

The ecosystem concept formed part of this dynamic: defined in terms of the flows of matter and energy, the concept was general enough to encompass natural communities otherwise distinguishable in terms of dominant species or other distinctive features. Thus, the concept was not only a particular set of ideas about nature but also a strategy contributing to formation of national communities of ecologists, independent of their local contexts. This strategy was especially expressed in the work of Tansley, Odum, and other ecologists, who saw the ecosystem concept as a basis for a unified ecology. Its independence from local context was also perhaps what one eminent ecologist, Pierre Dansereau, was thinking of when he described the study of ecosystems as that part of ecology "most nearly self-sufficient in its formulation of hypothesis and its choice of means to test them."[13]

National institutions reinforced both dimensions of this growth: in numbers of ecologists and in their sense of an identity transcending local context. For example, the National Science Foundation and the Atomic Energy Commission (AEC)—the two most important sources of funding for postwar American ecology—both discouraged research of mainly local application, viewing it as incompatible with their mission of supporting basic research. Similar priorities were evident in much of the support given to research at the Nature Conservancy, and in Ontario.

Before 1940 the balancing of responsiveness to local priorities against the building of their discipline had been a persistent challenge to ecologists. Many ecologists, particularly in the United States, relied on institutions stressing local relevance, such as agricultural experiment stations or state natural history surveys. Universities seeking funding and students also had to be responsive to local interest groups. In some cases, success in balancing local relevance and disciplinary objectives spelled the difference between a sustained research program and one with perpetually uncertain prospects.[14] After World War II this balance had become a much less pressing consideration in the building of ecological research programs. .

In this context, the perception of the environment as a political issue was in fact not new to ecologists but a return, to some extent, to once-familiar challenges. Like the concerns that had once justified ecological research—agricultural pests, choice of crops, fish yield, or protection of valued habitats, for example—many contemporary environmental controversies originate in concern about a specific local issue or place. The expertise they demand, accordingly, must also often be relevant to a specific context. A difference, however, in this challenge, as experienced by ecologists earlier and in the late 1960s, is that in the latter case, ecologists encountered it after several decades of building national research communities, within which they had not had to balance local issues and disciplinary priorities (or, in the Nature Conservancy, had been able to define this balance themselves).

In consequence, many ecologists, particularly those seeking to contribute to public affairs, experienced conflicting pressures: those imposed by their discipline and by the constraints of environmental politics. This was sometimes confusing. As George Woodwell noted in 1976, "The [Ecological Society of America] seems in disarray when presented with worldly problems. It is uncertain, torn between traditional objectives of scholarship and progressively consuming participation in public affairs as the environmental squeeze tightens."[15] The contrast was partly, as I have noted, one of scale: between national or international research communities and local controversies (epitomized by the general models developed in the International Biological Program and the specific models used in the Hudson River controversy). It also had to do with different standards of scientific activity—those imposed by the scientific community, and those imposed by the legal system, government bureaucracies, and the public.

Ecologists responded to these conditions in a variety of ways. Some perceived their best choice to be methods—such as fish population models or simulation models of radionuclide movement—able to provide precise predictions in specific environments. Other ecologists chose to ignore the application of their research to specific situations, in favor of contributing to a theoretical understanding of some aspect of ecosystems. To some extent, this compromise was unavoidable. As Richard Levins explained in the 1960s, a model cannot be at the same time general (applicable to all situations), realistic (representing the actual processes occurring in nature), and precise (providing quantitatively exact predictions). Instead, it is necessary to sacrifice generality for realism and precision, give up realism in favor of generality and precision, or aim for realism and generality, but not precision.[16] However, as we will see, some ecologists also defined new roles for their expertise, combining research of interest to ecologists generally with relevance to environmental issues.

To understand these choices it is necessary to see them as responses to different opportunities and constraints presented to ecologists by and through their institutions. In the next section I begin a comparative examination of ecology at each of the four institutions. My initial focus will be on the time when the environment was not a prominent political issue—that is, before the "environmental revolution."[17]

BEFORE THE REVOLUTION

As we have seen, Nature Conservancy ecologists emphasized the responsibility of their institution for both research and nature conservation. They considered this dual mandate to be essential, not only because conservation required research but also because it facilitated the use of reserves for research and ensured that ecology would benefit from practical concerns, just as Tansley had been arguing for two decades. This mandate and other features of the conservancy, including its support for long-term, collaborative research, also enabled ecologists to pursue new research approaches, such as production ecology and ecosystem ecology, seen as essential to the development of their discipline. Stressing the conservancy's distinctive mandate, ecologists also distinguished their work from that of agriculture and forestry, asserting its fundamental character and its concern with long-term problems and with marginal land neglected by forestry or agricultural research.

The Nature Conservancy maintained two research efforts, one focusing on specific, local tasks, such as the management of nature reserves, the other devoted to matters of more theoretical interest, such as the movement of nutrients within forest ecosystems. Both reflected ecologists' view of how they could best contribute to British environmental politics. By managing their own reserves and by working with local interests, such as naturalists, landowners, or farmers, they could infuse ecological insights and priorities directly into land management practices. A theoretical understanding of forest ecology could also provide a credible basis for contributing to the work of such national agencies as the Forestry Commission. Overall, the Nature Conservancy, in its research, advisory, and reserve management activities, epitomized the concern of such ecologists as Nicholson, Tansley, and Pearsall for wildlife habitats in the British countryside. At the same time, it asserted their authority as the experts best able to protect and manage these habitats.

Each year its annual report provided a lengthy account of the issues to which the conservancy had contributed advice. Nicholson and his colleagues believed that cooperation with other interests, with the conservancy providing scientific leadership, enhanced its effectiveness: "It has always been the Conservancy's view that conservation is much too great a task to be done by the Conservancy alone. They can merely study it, experiment with it, demonstrate the need and how to meet it, and act as a pace-setter, guide and stimulus towards a nation-wide understanding and fulfillment of this great task."[18] Indeed, cooperation was the guiding principle in the transfer of ecological knowledge to practical issues. It was facilitated by the conservancy's regional organization, staffed by officers that understood both the local landscape and its people.

Reliance on cooperation reflected an assumption prevalent among conservancy ecologists, and particularly Nicholson: that with effective communication and a scientific understanding of the issues, agriculture and forestry (the most important uses of the countryside) would not conflict with conservation objectives.[19] Cooperation was also consistent with the reliance on "consensus, compromise and gentlemanly agreement" characteristic of land use management, and government-private sector relations generally, in Britain. Since the 1840s British regulators had eschewed confrontation, pursuing environmental goals through corporatist relationships of consultation and negotiation with private interests. The Nature Conservancy therefore exemplified these views concerning the appropriate conduct of government, and its relations with society.[20]

Concern about radioactive contamination from nuclear waste provided the first opportunity for ecology to take root at Oak Ridge, and subsequent studies by Auerbach and his colleagues occasionally addressed other environmental problems, including the potential hazards of nuclear projects, such as the Plowshares program. They also asserted the relevance of their work to certain nonnuclear issues, consistent with Weinberg's intention that his laboratory address the largest, most complex problems of society. The Walker Branch watershed study exemplified such efforts. The significance to ecology of these problems, and of the managerial perspectives prevalent at the Oak Ridge laboratory, was evident in the use of innovative research methods, including radiotracers, simulation models, and collaborative efforts involving numerous scientists. Practical relevance helped justify these innovations, while they contributed to the reputation of Oak Ridge in radioecology and ecosystem ecology.

But, as I have suggested, ecologists were rarely called on to address actual environmental problems, both because the AEC considered these problems to be of low priority and because other forms of expertise were often considered more relevant. Thus, radioecology and ecosystem ecology flourished at Oak Ridge not by providing specific empirical predictions or advice concerning the environmental aspects of nuclear energy, but rather through support for the AEC's demonstration of the contribution of its technology to American society, and to national security. This reflected the political context of nuclear energy in the United States. During the 1950s and 1960s, as I noted in chapter 3, responsibility for nuclear energy was held by an iron triangle of government and industry interests: the AEC, its legislative overseers, and the nuclear industry. Debates on nuclear issues occurred largely within this corporatist structure, among officials sharing a broad consensus on the promise of nuclear power and the manageability of its risks.[21] The demand within this structure was not for critiques of nuclear power or specific advice concerning its risks, but for research supportive of the overall objective of furthering nuclear technology. Ecology gained a place at Oak Ridge by remaining consistent with this prevailing consensus within the American nuclear establishment.

A distinctive feature of the Hubbard Brook study was its connection between ecology and forestry. This was evident in several aspects of the study: the experimental manipulation of watersheds and metering of water flows were methods adapted from Forest Service studies; forestry research elsewhere corroborated results from Hubbard Brook; specific insights, such as the significance of erosion, were also derived from forestry research. For Bormann and Likens this connection served both scientific and environmental goals. Scientifically, it facilitated a new approach to ecology, based on experimentation, and analysis of the physical and chemical properties of ecosystems, through which they could colonize the no-man's land between established disciplines concerned with the soil, water, or biotic components of ecosystems. Environmentally, this approach would provide the integrated perspective necessary for managing the impacts of human activities on the landscape.

Nevertheless, the different objectives of ecologists and forestry researchers were also evident. While ecologists sought a general understanding of the behavior of intact forest ecosystems, forestry researchers were interested in the environmental impacts of specific forest cutting practices. This distinction, evident in different experimental approaches, reflected the structure of the study as a combination of university researchers funded by an agency—the National Science Foundation—that had no mandate to be involved in resource management or environmental politics (indeed, as Bormann and Likens would find, that would resist involvement in specific issues); and of Forest Service researchers, who had a responsibility to address their agency's resource management priorities.

But the ecologists and Forest Service researchers did share a belief (originating in the progressive conservation era) in professional scientific expertise as the necessary basis for natural resource management. This implied that resource management could be most effectively furthered not through controversy but through cooperation and consultation among professionals, as epitomized by the close, mutually beneficial ties that Bormann and Likens formed with Pierce and other Forest Service researchers at Hubbard Brook.

The ties between ecology and resource management evident at Hubbard Brook were tighter, if anything, in both institutional and intellectual terms, at the University of Toronto and the Ontario government. Ecologists addressed both scientific and resource management goals. Their role of providing a scientific basis for fisheries management meant steady support for research on fish population dynamics and environmental physiology. Their research also helped justify the managerial perspective dominant in Ontario fisheries and, indeed, in Canadian natural resource management between the 1940s and 1960s, that emphasized full exploitation of resources.

As the province developed its own corps of expertise, government biologists addressed the diverse specific issues that arose throughout the province. University ecologists, in contrast, while supporting the objectives of fisheries management, generally focused on studies at a few sites, seeking a basic understanding of fish populations and their physiology. These distinct objectives reflected the institutional structure of fisheries research in Ontario: the growth of government scientific activity; the independence accorded university scientists; their preference for fundamental research; and their importance, as scientists from the largest provincial university, in providing a rationale for provincial policies.

As in Canadian resource management generally, this reliance on expertise in fisheries management, and, in particular, the goal in fisheries research of a scientific basis for maximum yield, reflected the conception of resource management as primarily a technical problem, susceptible to solution through research focused on specific resources. As events in the 1970s would demonstrate, this view would eventually demand extensive revision.

Ecology, like science generally, does not exist in isolation but is embedded in and shaped by its context. An important part of this context consists of other disciplines. In these episodes, other disciplines have influenced choices that ecologists have made, or were obliged to make, in their research. Brief comparison of three cases will illustrate this.

Before the Nature Conservancy was created, British ecologists had been largely stymied in their efforts to obtain support for their research from such institutions as the Agricultural Research Council and the Forestry Commission, which were devoted to other disciplines. The absence of an institution devoted to ecology hindered their efforts to develop a distinct identity. The independent conservancy, in contrast, gave ecologists a powerful tool with which to develop this identity, distinctive both scientifically and in terms of its perspective on environmental issues. The political imperative of avoiding duplication of forestry and agricultural research, in the interests of government efficiency, reinforced this distinctiveness.

In contrast, ecology at Oak Ridge did not begin with an independent identity. Rather, it had a subordinate position in relation to the physical sciences, and specifically to health physics, which had originally justified its existence at the laboratory and provided most of its initial methods and research problems. This dependence was also evident in health physicists' view that ecology was merely a branch of their discipline. However, Auerbach and his colleagues did gain greater independence, both institutionally, in the form of a separate Radiation Ecology Section, and intellectually, with distinctive problems and methods. As they did so, they also asserted their distinctiveness in relation to neighboring disciplines, including health physics and biology. The ecosystem concept was a major element in this assertion.

Ecological research at Hubbard Brook did not begin, as it had at Oak Ridge, intellectually and institutionally dependent on another discipline. Nevertheless, Bormann and Likens linked their research with neighboring disciplines, including forestry and limnology, taking methods and ideas from both to use in developing a new approach to ecology.

We can see, therefore, how ecology's relationships with other disciplines have been intrinsic to its own identity. These relationships have occurred because from time to time ecology has had to solve certain practical or scientific problems that could not be addressed solely by tools available within the discipline. Some of these problems have stemmed from the mandates of the institutions supporting ecology. One example is the problem of tracing radionuclide contamination in the environment. Other problems have stemmed from ecologists' ambitions to revise their own discipline, using ideas from other disciplines (such as the notion of measuring physical and chemical changes within a forested watershed).

These accounts also demonstrate how institutional relations between disciplines can imply intellectual relations. At Oak Ridge, for example, ecologists initially depended on health physics both for their existence at the laboratory and for methods and research problems. As Jonathan Harwood has observed, and as other studies have corroborated, "To be institutionally dependent on another discipline is to be intellectually dependent as well."[22] Such relationships of dependency can stem from the differing scientific status of disciplines. Ecology has often been perceived, by ecologists and others, as having relatively little status, not being quantitative, rigorous, or simply "scientific" enough in relation to the standard set by physics. As a result, ecologists have often sought to raise their status by drawing on other disciplines having a more desirable status. One ambition shared by many ecologists at these institutions was to reshape their discipline, breaking from prewar "classical" ecology. This was expressed in different ways, reflecting different combinations of expertise, and varying demands of their institutions. Common elements included a focus on the functioning of ecological systems, as well as their structure; new methods and equipment, from radioisotopes to computer models to sophisticated chemical analysis; more quantification; and larger collaborative projects.

But while drawing on other disciplines to reshape ecology, many ecologists also affirmed the distinctiveness of their discipline. One way was by emphasizing experimental field study. The Nature Conservancy developed experimental nature reserves as "open-air laboratories." Auerbach described the Oak Ridge landscape as a research tool analogous to the experimental apparatus of the physical sciences. Hubbard Brook ecologists sought to "apply the methods of the experimental laboratory to the field." Ecologists in Ontario developed experimental fisheries and "experimental limnology."[23] These projects reflected a common interest in transferring experimental techniques to the field in order to create, in effect, outdoor laboratories for research unlike that undertaken by any other discipline. And for many ecologists, as we have seen, the ecosystem concept brought both strands together, enhancing the scientific status of ecology while asserting its distinctive identity as a discipline with a unique perspective on the world.

We have also seen how environmental issues played significant roles in these efforts to assert the status and distinctiveness of ecology, providing motivation, opportunities, or techniques for innovations in research. Prior to the environmental revolution, ecologists were not usually compelled to address environmental issues. They nevertheless often chose to, either because of their own concerns or because doing so furthered their scientific ambitions. In effect, environmental issues served as resources for ecologists, assisting them in building their discipline in cognitive terms and in terms of their professional identity.

That ecologists were able to redefine environmental issues in this way reflected the political context of environmental and resource issues in Britain, the United States, and Canada between the 1940s and 1960s. Within this context, responsibility for these issues (including nature conservation, atomic energy, forestry, and fisheries) was restricted to limited sets of public and private agencies, which shared certain values and assumptions and were insulated from outside scrutiny. Such arrangements have been characterized as iron triangles, or, more generally, as corporatist structures of governance. Their significance was enhanced by the relative absence of concern about these issues outside these structures. In Britain, nature reserves and wildlife conservation had not attracted much concern outside of naturalist, ornithological, and similar societies. In the United States, concern about the environmental implications of nuclear power was not yet widely evident; nor was there substantial public debate about forest management. And in Ontario, the condition of the Great Lakes had not yet attracted widespread concern.

An additional feature of these corporatist structures were certain assumptions about science. After World War II, scientists experienced unprecedented authority, with numerous government responsibilities—from conservation to development of new energy sources—defined as amenable to solution through scientific expertise. The scientific communities of Britain, the United States, and Canada also enjoyed considerable autonomy in establishing research priorities.[24] Ecologists at each of the institutions I have considered shared in this autonomy.

To sum up, environmental issues were defined not by society as a whole but within these corporatist structures. The situation was similar with respect to scientific priorities. In consequence, ecologists were often able to define these issues (particularly when other, more authoritative disciplines did not have a compelling interest in them) in ways that, by addressing them, they could also address their scientific priority: the development of ecological methods and theory.

The debate on social relevance—between a sense of responsibility to participate in public affairs and a preference to avoid distraction from scientific priorities—has engaged ecologists for decades.[25] However, the corporatist political contexts of the institutions I have discussed blunted this debate. In effect, a strict trade-off was not really necessary. Accordingly, many ecologists were receptive to concepts, like that of the ecosystem, especially as presented in Odum's *Fundamentals of Ecology*, that combined social and scientific relevance. It was within this context, then, that when environmental issues became prominent in the late 1960s, many ecologists expected to play a major role in addressing them. In fact, the environmental revolution entailed more than just greater prominence for these issues; it generated a variety of consequences for ecologists, and for their role in environmental politics.

ECOLOGY AND THE ENVIRONMENTAL REVOLUTION

While the environmental revolution drew from sources outside the discipline of ecology—both specific works, such as Rachel Carson's *Silent Spring*, and larger shifts in social attitudes—many nevertheless recognized ecological expertise as relevant to the new environmental priorities. The Nature Conservancy's advocacy, and that of prominent naturalists like Frank Fraser Darling, helped ensure this in Britain. In Canada, the perceived relevance of ecology was reflected in funding for the International Biological Program and by ecological initiatives of the Fisheries Research Board, particularly at its new Freshwater Institute in Winnipeg.[26] In the United States, Stewart Udall, Secretary of the Interior under Presidents Kennedy and Johnson, used to ask "What's ecology?" By 1964 he was predicting that the 1960s would become known as the Era of Ecology, and he urged ecologists to speak out on environmental issues.[27] Four years later, Congress supported the International Biological Program, citing a pressing need for ecology: "As the hearings progressed . . . it became ever more clear to the subcommittee that the IBP . . . dealt . . . with one of the most crucial situations to face this or any other civilization—the immediate or near potential of man to damage, perhaps beyond repair, the ecological system of the planet on which all life depends."[28]

These official perspectives mirrored more popular perceptions of ecology as a way of making sense of a degraded environment: "Ecology was the science which could interpret the fragments of evidence that told us something was wrong with the world—dead birds, oil in the sea, poisoned crops, the population explosion. What it meant was—everything links up. Here was a new morality, and a strategy for human survival rolled into one."[29] As this declaration suggests, some, particularly in the United States, saw ecology not merely as another form of specialized expertise but as a synthetic perspective able to, as Frederic Clements had once suggested, restore the unity of scientific understanding of the environment.[30] By 1966 this view of ecology was informing political debate. The Congressional Subcommittee on Science, Research, and Development stated that year that "ecology, as an organized profession, is not in good condition to become the umbrella for increased research. As a scientific discipline it is the logical focal point. As a point of view it is already effective in coordinating other sciences, and this may be the most important function in the long run."[31]

This view of ecology also fit well with the notion that various social problems, from poor public health to urban decline, were the result of piecemeal, uncoordinated decisions that addressed special interests but neglected the public interest. As one report noted, "Many of our most serious problems have arisen because narrowly conceived technological improvements have failed to take account of side effects."[32] Lynton Caldwell identified a divergence between ecological realities and the "tendency to focus upon one problem at a time[,] . . . to act upon the environment in a variety of apparently unrelated ways toward unrelated or unreconciled objectives. The human environment is a single, complex ecological entity, but man persists in acting upon it as if each of his actions had no bearing upon any other aspect of the environment."[33] Targets of similar critiques included established institutions, such as the Forest Service (and its preoccupation with timber production) and the Atomic Energy Commission (and its focus on a single form of energy).[34]

Such critiques remind us that in all three countries the environmental revolution entailed more than just greater concern about pollution and nature. It also signified a rejection of narrow regulatory and promotional practices engaged in by government. Instead, consideration of wider ranges of policy alternatives was demanded. This was reflected in several initiatives: the formation of the Natural Environment Research Council in 1965 from the Nature Conservancy and several other agencies; the creation in 1971 of Environment Canada, a new federal agency that some expected would adopt a comprehensive approach to environmental and resource management; and the establishment of the Environmental Protection Agency in 1970, and of the Energy Research and Development Administration (subsequently the Department of Energy) in 1974.[35]

Confidence in comprehensive perspectives as a way to deal with complex problems was also reflected, especially in the United States, in the widespread adoption of systems analysis and other management strategies.[36] A few successful projects, like the Apollo space program, had suggested that national goals could be achieved through the large-scale marshaling of technical expertise. If a man could be put on the moon, surely comprehensive, technically-based strategies, backed by a strong, activist government, could address other national goals, including a clean environment. Such strategies would encompass all aspects of nature and society, a broader range of policy alternatives, and the technocratic conviction that experts could define and achieve optimal goals for society.

Reliance on comprehensive perspectives depended on certain assumptions: that environmental problems demand an urgent response, that the government should adopt a strong role in addressing these problems, and that scientific expertise should lead in identifying problems and defining solutions. All these assumptions were questioned in the 1970s, particularly in the United States. A consensus on the urgency of environmental protection, briefly present during the late 1960s, had dissipated. Confidence in the capability of government to address the public interest was eroded by Watergate, the Vietnam War, and economic uncertainty. Scientific controversies and uncertainties lessened the authority of science as a source of certain, optimal solutions to society's problems. Environmental activists demanded more open, democratic participation in government decisions. In American environmental politics, accordingly, the rational comprehensive approach declined in authority. It was replaced by a more pluralistic regulatory process that relied substantially on judicial or quasi-judicial mechanisms and within which a wider range of interests participate. These interests usually do not share basic values and goals, as interests within corporatist structures had. The result, often, has been greater confrontation between opposing interests.

Scholars of comparative policy generally agree that environmental politics are more adversarial in the United States than in the other countries. In particular, the British have continued to rely more on cooperation between regulators and the regulated, with less frequent recourse to the legal system. However, some British environmental groups have, more recently, adopted confrontational strategies as the basic conflict between conservation and modern agriculture and forestry has become increasingly evident.[37] Canadian environmental politics also appear to be less adversarial than that of its neighbor: there has been less recourse to litigation in environmental controversies, and regulatory agencies have continued to rely to a large extent on cooperation with those they regulate.[38]

Another feature of the evolving political context of ecology after the environmental revolution was a "new politics of science," in which research priorities would be defined outside the scientific community. No longer would it be assumed that science can best benefit society by being left on its own; instead, it would be directed toward national priorities. In Britain, interest in linking science to practical concerns had been building since the late 1950s; this culminated in 1972 in a "customer-contractor" approach to setting scientific priorities, which gave government departments more control over research councils. In Canada, a formal science policy and planning apparatus had begun to emerge by 1970 as successive studies asserted the need for mechanisms to direct science and ensure its political accountability. In the United States, Presidents Johnson and Nixon attempted to mobilize research toward national priorities, and congressional committees examined the need for an effective science policy.[39]

The practical consequence of these developments was that the politics of science, like environmental politics, became more pluralistic: more public agencies and private interests began to support research addressing their priorities. These trends came together, in that in environmental politics a major role for scientific expertise became that of providing a basis for the confrontation, negotiation, and brokerage of competing interests, as well as support for assertions of the rationality of administrative agency decisions and policies.[40] Whether in environmental impact assessments or in the larger conflict between a clean environment and the cost of achieving it, the demand has often been for precise information that can serve as a basis for negotiation between adversaries. This demand has been reinforced by the persistent tendency to define environmental problems as amenable to technical solutions, thereby reducing the need to deal directly with conflicting values and interests.

In the late 1960s and early 1970s, therefore, several circumstances defined the new politics of the environment: widespread environmental concerns; a perception that ecology was relevant to these concerns; a retreat from corporatist structures of environmental and resource management, replaced by demand for comprehensive strategies for addressing environmental issues; a subsequent decline in authority of these strategies, along with greater interest in directing science toward specific political and economic priorities; and finally, greater participation by a wider range of often opposing interests. Ecologists' responses to these circumstances can be interpreted in terms of their institutional contexts.[41]

The Oak Ridge laboratory had a variety of mandates. One, exemplified by participation in the International Biological Program, was to provide a basis for comprehensive approaches to environmental management. In addressing this mandate, Oak Ridge ecologists expanded their efforts in systems analysis and ecosystem modeling. This included efforts to apply models to practical issues. But, as we have seen, the more important development was the partial convergence of Oak Ridge ecology and academic ecologists, as reflected in their development of ecosystem theory, along with a distancing, to some extent, of this research from much of environmental politics.

The evolution of environmental politics and of the laboratory—new regulations, responsibility for a broader range of energy sources, demands for technical expertise from other agencies, ecologists' efforts to seek additional clients—fostered other mandates. Pluralistic, adversarial environmental controversies generated sharper critique of scientific evidence than was customary within the scientific community. As Oak Ridge ecologists found in the Hudson River controversy, the demand was for advice that would be perceived as factual and objective, that could provide certainty of cause and effect, and that could withstand examination and sharp challenge by opposing interests. These demands discouraged the use of ecosystem theory and models in favor of representations of fish populations better able to provide quantitative predictions.

American ecologists would often find that ecosystem ecology could not fulfill the requirements of regulatory agencies.[42] The demand was not for comprehensive understanding of environmental issues, but for studies on specific immediate problems. In 1980 one observer noted that within the Environmental Protection Agency the "demands of justifying and monitoring regulatory procedures tend to take precedence over holistic approaches, and the contracts and grants follow suit. Lawyers operating within statutory compartments and scientists with circumscribed expertise still tend to dictate the immediate study agenda."[43]

The displacement of ecology at Oak Ridge from its primary role in environmental research—symbolized by the renaming of the Ecological Sciences Division as the Environmental Sciences Division in 1972—can also be understood within the context of the demands on expertise imposed by American environmental politics. Ecologists found that many environmental controversies—from wilderness protection to urban air quality—demanded expertise in disciplines other than ecology. For example, one of the first controversies over reactor siting, at Bodega Bay, California, focused on scenic landscape values and potential seismic activity; neither issue could be addressed by ecologists.[44] In general, environmental controversies have usually been defined in terms relevant to human experience and values; as a result, they tended to focus not on the overall condition of ecological systems but on aspects of particular aesthetic, recreational, or economic significance. As one ecologist has noted, "The ecologist sees himself as concerned with the general health of an ecosystem, while all too often the patrons of ecology . . . focus only on a specific unpleasant symptom."[45] This was especially evident in the shift during the 1970s in concern about environmental contaminants, from their potential impact on natural ecosystems to their risks to human health. The Toxic Substances Control Act of 1976 and related regulations generated demands for expertise in environmental toxicology and epidemiology, and the Environmental Protection Agency focused increasingly on the protection of human health.

With the loss in the early 1970s of consensus over the urgency of environmental protection, economists and other social scientists also became more important. No longer was the environment to be protected regardless of cost; instead, cost-benefit analysis, risk assessment, and other methods of establishing relative priorities became important.[46] These methods required that the benefits of environmental protection be translated into economic terms.

As a result, environmental analyses tended to focus on damage to property and to human health, for which costs can be calculated. Noneconomic values, such as species of no commercial significance, were frequently excluded. These developments also lessened demand for ecological expertise.

Technological expertise also became more prominent. Environmentalists, inspired by E. F. Schumacher, Amory Lovins, and other writers, pressed harder for better technologies—such as photovoltaic cells, more efficient engines, and similar innovations—not for better understanding of ecological systems.[47] Environmental regulations have also contributed. The 1972 Clean Water Act, for example, established the goal of restoring the "integrity" of the nation's waters. While integrity was to be defined in ecological terms, the act aimed to achieve it by requiring each category of industry to adopt the best available treatment technology. The purpose of this requirement was to avoid, when prosecuting violators, the need to prove damage to the environment. This requirement in earlier statutes had resulted in long legal disputes because it was difficult to establish with certainty that any one act of pollution had caused damage. On the other hand, it was expected to be easier to establish whether the technology in use was not the best available. This illustrates how the courts, in seeking certainty of cause and effect, will prefer the least uncertain scientific evidence.[48] But because of the complexity of natural systems, as well as conceptual difficulties and controversies within ecology, advice provided by ecologists can often be highly uncertain.[49] Inevitably, therefore, those who write environmental laws will often avoid relying on ecological evidence, in order to make them less susceptible to challenge.

Several specific factors, then, contributed to the displacement of ecology from a primary role in environmental politics at Oak Ridge. This also reflected a change in the institutions of environmental politics. When the AEC had provided almost all the funding for Oak Ridge ecology, Auerbach, Odum, Wolfe, and other AEC ecologists had been able to define the environmental implications of nuclear power so as to be consistent with their ambitions for radioecology and ecosystem ecology. By the end of the 1970s this was no longer possible. Many more agencies had become responsible for environmental issues, and with the consequent shift in control over environmental research to these agencies, Oak Ridge ecologists had much less ability to define these issues in terms of their own expertise.

Oak Ridge has nevertheless continued to be an important center for ecological research. For example, a large program in fish ecology has continued since the early 1970s, to support involvement in regulatory activities like those on the Hudson River. To be sure, numerous environmental issues, from protection of endangered species to restoration of damaged ecosystems, have continued to demand ecological expertise. Ecosystem ecology has also continued as an active area of research. Much of this research, as I have noted, has moved closer to the theoretical priorities of the academic research community. However, ecosystem ecologists have also addressed certain environmental issues. They, and their colleagues elsewhere, have argued that their research, by providing a theoretical understanding of natural systems, is a more effective basis for environmental protection.[50] Environmental management, for example, would be more effective if based on knowledge of the factors influencing ecosystem stability and resilience. Ecosystem ecologists have also suggested that their research can anticipate emerging environmental problems before they become evident to the public or policymakers. For example, studies of climate change as a result of increasing carbon dioxide in the atmosphere began long before there was public awareness of this problem. Scientists have repeatedly stressed the importance of "anticipatory" research in identifying problems before they become irreversible.[51]

More generally, many scientists have argued that ecosystem research, and basic research generally on the environment, have an essential function as an alternative to the short-term instrumental priorities of regulatory and management agencies. Another key feature of this "anticipatory" research is that in focusing on problems that have not yet entered the political arena they can be more readily redefined in terms of basic research priorities. For example, while concern about climate change has motivated studies of carbon dioxide, it has been argued that these studies should focus on the "interactions among the cycles of carbon, nitrogen, phosphorus, and sulfur. This alternative promises greater returns in the long run by attempting to understand the cycles themselves instead of trying to solve the CO_2 problem."[52] In addition, as such research proceeds largely within the scientific community, debate is less adversarial and more accommodating of uncertainty than is typically the case when political and economic interests are in play. This permits greater consideration of the imprecise, tentative results characteristic of the study of complex systems, whether ecosystems or the entire biosphere. In summary, therefore, while ecology was displaced from a preeminent role in environmental politics by other forms of expertise better able to address its varied demands, some ecologists were able to construct research perspectives combining a distinctive contribution to environmental issues, with attention to theoretical concerns. This effort, it must be noted, has been dependent on agencies, such as the NSF, willing to support research not addressing specific environmental controversies.

Hubbard Brook ecologists encountered the new environmental politics directly, when their results were invoked in support of a denunciation of forest cutting practices before a congressional committee. This episode, and the national controversy within which it occurred, reflected the adversarial character of American environmental politics in the 1970s—a character we have already encountered in the Hudson River episode. This controversy also signaled a growing resistance to the management practices of the U.S. Forest Service and, more generally, to the notion of professional expertise as the sole determinant of resource management practices.

As we have seen, Bormann and Likens chose not to become involved in this controversy. Instead, during the 1970s they strengthened their ties with forestry managers and researchers, advising them on the ecological implications of forest harvesting and receiving in return ideas, methods, and data. Through publications and workshops and other forms of direct communication they participated in discussions among experts concerning the technical aspects of forest management.[53] They also contributed to public discussion of forestry issues. With American forests now of concern to a wider range of interests, a broad-based discussion about them began: their values when maintained intact, their appropriate uses, and their vulnerability to human impacts. Within this discussion Bormann and Likens explained their results in books and their less technical articles. By warning of such impending problems as soil nutrient depletion or acid precipitation, Bormann and Likens fulfilled (like some Oak Ridge ecologists) the "anticipatory" function of some environmental research. In addition, by influencing public attitudes about the values and vulnerabilities of forests they epitomized the scientific role that political scientists have described as "enlightenment": the gradual development of public understanding of the scientific aspects of environmental issues.[54]

As I have already noted, ecologists and foresters at Hubbard Brook had different research objectives. This difference was the key to the ecologists' distinctive roles in public debates about forestry: that they were defined not in terms of specific problems identified by foresters but in terms of their distinctive ecological perspective, based on the ecosystem concept. They considered their synthetic ecosystem perspective essential to communicating the ecological complexities of forests and to linking these complexities with the range of views represented in the debate. Their ability to maintain their distinctive ecosystem perspective reflected their institutional status. As university scientists funded by the National Science Foundation, they had the latitude to define their research in ways that would address their professional and disciplinary goals, as well as their own concerns about forests.

During the 1970s, however, an important component of Hubbard Brook research became more remote from environmental issues. At the start of the decade, students had often explored in their papers the practical implications of their results. By 1980, however, this had become much less common, as they focused more strictly on scientific issues. This mirrored changes within the discipline. As has often been noted, relevance to environmental concerns became less important to the self-image of ecology during the decade, even as ecology was itself becoming less prominent within environmental politics. Students responded more readily to this than did Bormann and Likens. This may have been because as they prepared to enter an increasingly competitive academic job market they had to be sensitive to changing priorities within their discipline. A capacity to consider the societal implications of their research had become less of a selling-point to prospective university employers. In contrast, Bormann and Likens, as more senior scientists, were able to define their own relation to public issues, even challenging, as we have seen, the National Science Foundation's view on this matter.[55] This distinction between junior and senior members of the study enhanced the importance of Bormann and Likens in synthesizing the results of their study and establishing its practical implications.

Until the early 1970s, ecologists at the University of Toronto had defined their relation to fisheries management primarily in terms of achieving maximum sustained yield. Eventually, however, this contribution could itself no longer be sustained. Evolving public priorities, including a wider range of values attached to the aquatic environment, as well as a degraded environment—impoverished fish populations and diminished habitats—inconsistent with these new values, together made imperative a new approach to aquatic environmental management. At the same time, university ecologists were developing new research strategies, focusing on aquatic ecosystems and the stresses affecting them. Ecologists, including Regier and his colleagues, defined their role within society in terms of the capability of these strategies to provide a basis for comparing and evaluating the stresses imposed by humans, directly and indirectly, on fish communities. During the 1970s they contributed to a new approach to fisheries management, the Strategic Plan for Ontario Fisheries, the goal of which was not maximum yield, but the maintenance or restoration of healthy fish communities. The political viability of this plan was enhanced by contemporary interest in comprehensive approaches to environmental problems. This paralleled a similar interest in the United States; however, it may have persisted longer in Canada, as Canadian environmental regulation has characteristically exhibited greater administrative discretion and less reliance on adversarial processes.[56] The strategic plan epitomized these features in its application of a comprehensive ecosystem approach to managing fish communities, intended to reduce resource use conflicts.

The role of Canadian ecologists in this episode of fisheries management resembled that played by Bormann and Likens (and their Forest Service colleagues) in forestry management: the promotion of an ecosystem perspective on the use and abuse of environmental resources. However, while Hubbard Brook ecologists developed this role based on research begun in the early 1960s, the Canadian ecologists' contribution was based on what was, for the University of Toronto, a new approach to ecology. This approach reflected not only new ideas but also a changed institutional context. For many years scientists at the University of Toronto had had a close relationship with provincial resource management. But by the early 1970s this relationship had been eroded as the government came to reply more on its own expertise, and the scientists that had maintained it retired and were replaced by ecologists with interests less relevant to traditional management objectives. Their role in providing a critique of traditional management practices, as well as an alternative, was facilitated by them having acquired institutional independence from the management agency.

9

Understanding Ecology's Place in the World

Even as the need to understand and address environmental problems has become more acute, changes in society have made the role of ecology in humanity's relationship with the natural world more complex than ever. For historians seeking to understand this complexity, at least two strategies are available. One is to examine the dominant values of society, and the extent to which ecologists have exemplified or opposed these values. This strategy acknowledges that while ecology has told us much about the natural world, this knowledge reflects not only nature but also the cultures within which ecology is embedded. But this strategy also, as I have suggested, transcends the complexity of environmental politics, sometimes reducing it to a clash of ahistorical values expressed in terms of a preservationist-exploitationist or ecocentric-anthropocentric or some other dichotomy. The alternative strategy is to examine the specific, diverse contexts within which ecologists work, and to consider the choices they have made from the options presented to them by these contexts. This strategy embraces the complexity of environmental politics and the historical contingency of the relationship between ecology and social values.[1]

For ecologists, the challenge of combining scientific and social priorities—of asserting an ecological perspective on social priorities, and simultaneously, their authority as scientists able to address these priorities—has become much more formidable in the last three decades. This has been reflected in the decision of many of them to focus on objectives defined within their own discipline, even as other forms of expertise have responded more successfully to demands from government, business, and activists for environmental expertise. A chief lesson of the environmental revolution for ecologists was that they could not take for granted a central role in addressing environmental concerns.

Ecosystem ecologists, especially, had to learn this lesson. Studies of the history of the ecosystem concept tend to emphasize its basic ideas: the significance of flows of nutrients and energy, and of emergent properties of system regulation and control, for example. However, the concept was also a strategy: for uniting ecologists, for defining their distinctive perspective on the natural world, and for asserting ecologists' authority within environmental politics. The limitations of this strategy became evident during the 1970s and 1980s. While the division between population and ecosystem ecology persists, and prospects for a unified ecology remain distant, ecosystem ecology, particularly within American environmental politics, has often been unable to address specific public or administrative concerns, or provide predictions able to withstand adversarial challenge. Where it has been more effective is in those contexts—exemplified by the collegial community of scientists at Hubbard Brook, or by those seeking a comprehensive perspective on Great Lakes fishery management—within which other approaches have been perceived as inadequate, and that emphasize the search for consensus over the adversarial challenge of opposing interests. More speculatively, it may be anticipated that ecosystem ecology will play a more important role to the extent that consensual processes, within epistemic communities, or among wider arrays of interests, become more significant in environmental politics.[2]

The evolving status of the ecosystem concept within environmental politics illustrates how values and attitudes—concerning not only the natural world but also the role of expertise and government—have shaped the place of ecology within society. To understand how this has occurred, it is necessary to consider and compare the institutions that provide the immediate context of ecologists. Most obviously, certain institutions have provided ecology with the funds essential to its expansion since the 1940s. However, their significance was not only in terms of dollars and pounds. They have also added their own dynamic to the history of contemporary ecology. Sometimes they have fostered innovation: heavy support by the AEC during the 1950s, for example, spurred ecologists, at Oak Ridge and elsewhere, to take up the ideas presented by Eugene Odum and other early ecosystem ecologists. But they have also discouraged change: for example, the closeness of the ties between provincial fisheries management and ecologists in Ontario may have hindered application of new ecological insights until Great Lakes fish stocks had become severely depleted.

As we have seen in comparing the Nature Conservancy, Oak Ridge, Hubbard Brook, and the University of Toronto, these institutions embodied certain specific social priorities, such as support for science, environmental protection, technological development, or resource management. In addressing these priorities, they brought together particular combinations of expertise from the scientific community. In effect, these institutions served as mediators: between ecology and the diverse priorities of society, and between ecology and the rest of the scientific community. As mediators, they defined which priorities and disciplines would form the immediate context within which ecologists developed their research, effectively shaping the interaction of ideas and circumstance characteristic of contemporary ecology.

Environmentalism draws much of its strength from people's concerns about specific places or problems: as William Cronon has noted, there is no one big "environmental problem," but a near infinitude of smaller problems, that together reflect the diversity of ways in which people experience their environment.[3] So, too, will ecology—the work of ecologists of diverse experience and ambitions, influenced by the values and ideas of society and of other scientific disciplines, as mediated by an array of institutions—defeat efforts to assign it one "big" role in environmental politics, either in support of, or in opposition to, the exploitation of nature. Ecologists' encounters with nature and society have been more complex than this, and certainly more interesting.

Notes

CHAPTER 1. INTRODUCTION

1. See, e.g., Lubchenco et al., "Sustainable Biosphere Initiative," 371–412, Sears, "Subversive Subject," 11–13, and Odum, "Emergence of Ecology," 1289–1293.
2. Crowcroft, *Elton's Ecologists,* xiv.
3. O'Riordan, *Environmentalism,* ix.
4. Hays, *Beauty, Health, and Permanence,* 5.
5. Gottlieb, *Forcing the Spring,* 318.

CHAPTER 2. ORIGINS OF THE NATURE CONSERVANCY

1. See Hywel-Davies, Thom, and Bennett, *Britain's Nature Reserves,* 397–398; the inscription, written by Max Nicholson, is in *Report of the Nature Conservancy, for the year ended 30th September 1958,* 91.
2. Sheail, *Seventy-Five Years,* 39.
3. Tansley and Chipp, *Aims and Methods.*
4. Tansley, *Oxford Department of Botany,* 4–5.

5. Tansley, *British Islands*, x.

6. Tansley, *Oxford Department of Botany*, 5.

7. Salisbury, "Origin and Early Years," 16; Sheail, "Grassland Management," 297–298.

8. Stapledon, "Cocksfoot Grass," 103.

9. Crowcroft, *Elton's Ecologists*, 13–46; Sheail, "Applied Ecology," 68–70.

10. Robinson, "Ecological Aspects of Afforestation," 1–12.

11. British Ecological Society, "Ecological Principles," 83–87.

12. Tansley, *Heritage of Wild Nature*, 54; on the Forestry Commission plans see Forestry Commission, *Post-War Forest Policy*.

13. British Ecological Society, "Ecological Principles," 84, 88.

14. Champion, "Our New Forests," 319–324.

15. British Ecological Society, "Ecological Principles," 103. This was the focus of Watt's own research, which he would present in his influential paper "Pattern and Process."

16. Elton, "Population Interspersion," 1. See also Elton and Miller, "Ecological Survey," 460–496.

17. Hagen, *Entangled Bank*.

18. Tansley, "Use and Abuse," 284–307.

19. Tansley, "What Is Ecology?" 6–9.

20. Tansley, "Use and Abuse," 299.

21. Tansley, "What Is Ecology?" 15–16.

22. Elton, "Population Interspersion," 1.

23. British Ecological Society, "Ecological Principles," 86.

24. Pearsall, "Milestone," 241.

25. Salisbury, "Study of British Vegetation," 306.

26. British Ecological Society, "Ecological Principles," 103–104.

27. Tansley, "British Ecology," 527.

28. British Ecological Society, "Ecological Principles," 83.

29. Tansley, "British Ecology," 530.

30. Tansley's statement in meeting of the Advisory Council on Scientific Policy, November 2, 1949, PRO, CAB 132–66.

31. W. T. Thiselton-Dyer quoted in Allen, "Botanical Perspective," 208.

32. Sheail, *Nature in Trust*, 196–198; Harvey, "National Trust," 149–159.

33. Cherry, *Environmental Planning*, 13–15; Sheail, *Rural Conservation*.

34. Cherry, *Environmental Planning*, 33–37.

35. Nature Reserves Investigation Committee, "Nature Conservation in Great Britain," draft copy, January 1943, 1, PRO, HLG 92–18. (hereafter cited as NRIC, "Nature Conservation").

36. Society for the Promotion of Nature Reserves, "Nature Preservation in Post-War Reconstruction," conference memorandum no. 1, November 1941, PRO, HLG 92–18.

37. Tansley, *British Islands,* 192.

38. H. Smith to W. Jowitt, April 1, 1942, PRO, HLG 92–18.

39. NRIC, "Nature Conservation."

40. Tansley, "British Woodlands," 1–21.

41. Tansley, "Use and Abuse," 303, 304.

42. On Woodwalton Fen see Lowe and Goyder, *Environmental Politics,* 153. On Wicken Fen see Godwin, "Deflected Succession," 144–147, and Godwin, "Wicken Fen," 148–160.

43. This term was used at least as early as 1720; see Thomas, *Natural World,* 276.

44. Elton, "Land Ecology," 53; Graham, *Natural Principles.*

45. Huxley, *TVA: Adventure in Planning;* on the Wild Life Conservation Special Committee see report of conversations with Ira Gabrielson, U.S. Fish and Wildlife Service, November 25, 1945, and "Copy of a Memorandum of Agreement Between the National Park Service and the Fish and Wildlife Service, of the Department of the Interior, September 11, 1943," in Wild Life Conservation Special Committee, "Terms of Reference, Committee Papers, Minutes of Meetings, etc.," 1945–46, PRO, HLG 93–48.

46. Tansley, *Heritage of Wild Nature,* 35.

47. Committee on Land Utilisation in Rural Areas, *Report,* 4.

48. Ibid., 47.

49. The Nature Conservancy's slow progress during the 1950s and 1960s in developing practical techniques of reserve management and in training staff to apply them, confirmed that the ecologists' original, confident presentation of their ability to manage nature reserves had not been grounded in extensive experience or in full understanding of its practical implications. For this account, however, it is more significant that this presentation was persuasive in the 1940s.

50. British Ecological Society, *Nature Conservation,* 2.

51. Ibid., 32.

52. Sheail, *Seventy-Five Years,* 140.

53. Tansley, *Heritage of Wild Nature.*

54. Bukharin et al., *Science at the Crossroads;* Bernal, *Social Function of Science.*

55. McGucken, *Scientists, Society, and State;* Werskey, *Visible College.*

56. J. R. Baker, circular letter, November 2, 1940, reproduced in "History of the Foundation of the Society for Freedom in Science," in Society for Freedom in Science, *Origin, Objects and Constitution.*

57. Tansley, "Values of Science," 104–110.

58. Baker, *Arthur Tansley.*

59. Tansley, *Oxford Department of Botany,* 4 (emphasis in original), 17.

60. J. R. Baker, "List of Members of the Society for Freedom in Science on 12 July 1941," typescript.

61. Tansley, "[Discussion on Fundamental Research]," 294–295.

62. Tansley, *Heritage of Wild Nature,* 67.

63. Poole and Andrews, *Government of Science*, 157.

64. Society for Freedom in Science, "History of the Foundation of the Society for Freedom in Science," in *Origin, Objects and Constitution*.

65. Tansley, "Values of Science," 110.

66. Tansley, "What Is Ecology?" 10.

67. McGucken, "Freedom and Planning," 42–72.

68. Elton had also been convinced of the need for autonomy by his experience in the Bureau of Animal Population that the greatest progress is made by individuals free to engage in their own innovative research (Sheail, "Applied Ecology," 74–75).

69. Dower, *National Parks*.

70. J. B. Bowers to H. Smith, August 21, 1945, in National Parks Committee, "Correspondence with the Secretary of the NRIC (Dr. Herbert Smith)," 1945–46, PRO, HLG 93–39.

71. Sheail, "Nature Reserves," 29–34; Sheail, "Applied Ecology," 71–72.

72. Nicholson, *Birds in England*. David Allen discusses Nicholson's ornithological work in *Naturalist in Britain*, 252–258.

73. J. B. Bowers to H. Smith, August 21, 1945, H. Smith to J. B. Bowers, August 24, 1945, in PRO, HLG 93–39.

74. Minutes of first meeting of Wild Life Conservation Special Committee, September 6, 1945, in PRO, HLG 93–48.

75. Wild Life Conservation Special Committee, "A Suggested Pattern for the Administration of Wild Life Conservation and Nature Reserves in Relation to National Parks and National Research," November 9, 1945, 2, in PRO, HLG 93–48.

76. Wild Life Conservation Special Committee, "Draft Letter in Reply to Mr. Flett's Letter to Mr. Bowers," November 27, 1945, in PRO, HLG 93–48.

77. J. S. L. Gilmour, "The Use of Nature Reserves for Botanical Experiments," February 28, 1946, in PRO, HLG 93–48 (injecting a note of caution, Cyril Diver, in comments following this paper, urged that experiments in reserves be permitted only when their possible impact on existing communities was known); Wild Life Conservation Special Committee, "The Case for Experimental Bird Reserves," April 8, 1946, in PRO, HLG 93–29; Huxley's comment is in Wild Life Conservation Special Committee, "Note on a Joint Meeting with the National Parks Committee, February 20, 1946," in PRO, HLG 93–48.

78. Wild Life Conservation Special Committee, "Notes from Meeting of February 20 & 21, 1946," in PRO, HLG 93–48.

79. Morrison, "Planned Research," 296.

80. "Nature Preservation in England," 278.

81. Elton, "How a System of Nature Reserves Might Be Run," in PRO, HLG 93–48.

82. Wild Life Conservation Special Committee, "Note on Meeting with the Scientific Advisory Committee of the Cabinet," March 1, 1946, in PRO, HLG 93–48.

83. Sheail, *Nature in Trust*, 200–206.

84. Wild Life Conservation Special Committee, "Draft Letter in Reply to Mr. Flett's Letter to Mr. Bowers," November 27, 1945, in PRO, HLG 93–48.

85. Wild Life Conservation Special Committee, "Notes from Meeting of March 8, 1946," in PRO, HLG 93–48.

86. Sheail, *Nature in Trust*, 211.

87. Wild Life Conservation Special Committee, *Conservation of Nature*, 3.

88. Tansley, "Presidential Address," 195; Godwin, *Cambridge and Clare*, 150.

89. Elton, "Population Interspersion," 2.

90. Tansley, "British Ecology," 530.

91. Nicholson, *New Environmental Age*, 93; *Reports of the Nature Conservancy for the Period up to 30th September, 1952*, 1–3; Sheail, *Nature in Trust*, 215–216.

92. Morrison, *Government and Parliament*, 329–332.

93. Quoted in Sheail, *Seventy-Five Years*, 143.

CHAPTER 3. WOODLAND ECOSYSTEMS AND NATURE RESERVES

1. When the conservancy was only a few months old Nicholson told the Advisory Council on Scientific Policy that "unless the problem of finding additional manpower can soon be solved, it looks as if an almost crippling handicap is going to be placed on the necessary development of ecological work in general" (Memorandum, "Ecological Surveys," from Nicholson to ACSP, July 5, 1949, in PRO, CAB 132–66). Organizational problems of the early conservancy were reflected in scattered pressure for its abolition and by its failure to prepare an annual report during its first four years (Nicholson, E. Max., interview by author, London, May 15, 1990). Financial data are in the conservancy's annual reports.

2. In 1957–58, for example, the Department of Scientific and Industrial Research received £10,339,000, the Agricultural Research Council £4,626,000, the Medical Research Council £2,795,000, and the Nature Conservancy £304,000. On changing government priorities see Rose and Rose, *Science and Society*, 75.

3. Biographical details are in Pinder, *Fifty Years*, 214–215.

4. Under the Haldane principle, the councils operated almost entirely without governmental direction (Gummett and Price, "Central Planning," 120). Nicholson's comment is in Select Committee, *Nature Conservancy*, 67. In 1954 the annual report noted the "entire freedom which the Conservancy enjoy in their choice of researches and . . . the absence of any organised user to press the claims of particular projects" (*Report of the Nature Conservancy, for the year ended 30th September 1954*, 5). On the conservancy's reliance on the Scientific Policy Committee see Select Committee, *Nature Conservancy*, 8. Reinforcing the primacy of scientists in the conservancy, Nicholson once recalled telling its administrators that their "role was to absorb and implement the requirements which came from our largely scientific governing body, and were spelt out by the senior scientific staff, rather than to let their heads be troubled by policy formation or struggling on their own account for a more important role" (Nicholson, *New Environmental Age*, 96).

5. Select Committee, *Nature Conservancy*, 21.

6. *Report of the Nature Conservancy, for the year ended 30th September 1955*, 2. In discussions in the late 1950s and early 1960s over a proposed Natural Resources Research Council, the conservancy again argued that its autonomy be preserved (Poore, "Nature Conservancy and Nature Conservancy Council," 184).

7. "Interdepartmental Committee on the Nature Conservancy Draft Report," 1956, PRO, HLG 92–180. This committee included representatives of the Ministry of Agriculture, Fisheries, and Food, Scottish Office, Ministry of Housing and Local Government, Lord President's Office, and the Treasury.

8. Vogel, *National Styles of Regulation*.

9. *Report of The Nature Conservancy for the year ended 30th September 1953*, 12.

10. Worthington, *Ecological Century*.

11. Nature Conservancy, *Report, 1955*, 2.

12. Nicholson, E. Max., interview by author, London, May 15, 1990.

13. Nature Conservancy, *Report, 1955*, 2; Minutes of Meeting of the Nature Conservancy, July 8, 1955, in PRO, MAF 131–25.

14. "Interdepartmental Committee on the Nature Conservancy Draft Report," in HLG 92–180, PRO, 3. This committee had been appointed by the Lord President to study the conservancy and its relations to other departments; it included representatives from the Ministry of Agriculture, Fisheries, and Food, the Scottish Office, the Ministry of Housing and Local Government, the Lord President's Office, and the Treasury.

15. *Report of The Nature Conservancy, for the year ended 30th September 1964*, 5.

16. Nature Conservancy, *Reports, 1952*, 6–7; Sheail, *Pesticides and Nature Conservation*.

17. Wildlife Conservation Special Committee, *Conservation of Nature*, 62.

18. Biographical information is contained in Clapham, "William Harold Pearsall," 511–540. Pearsall was also active in the Freshwater Biological Association, and he was chairman of its council.

19. Ovington, "Study of Invasion," 35–52.

20. Ovington, "Culbin Sands," 303–319; Ovington, "Tentsmuir Sands," 363–375.

21. Tansley, "Use and Abuse," 284–307.

22. Tansley, "Early History," 131.

23. Pearsall, *Mountains and Moorlands;* Pearsall, "Soil Complex," 180–193, 194–205.

24. British Ecological Society, "Joint Meeting," 328.

25. Ovington, "Forest Floor," 71–80.

26. The work of each of these scientists is described in various reports of the conservancy.

27. Select Committee, *Nature Conservancy*, 55.

28. British Ecological Society, "Meetings, 1956," 610.

29. Ibid., 605.

30. Varley, "Experimental Science," 639–648.

31. Ovington, "Dry-Matter Production," 287–314; Ovington and Madgwick, "Growth and Composition," 271–283.

32. The British Ecological Society held a symposium in March 1956 devoted to production, which provides a survey of the extensive British interest in this field (British Ecological Society, "Meetings, 1956," 604–611; Pearsall, "Production Ecology," 106–111).

33. Pearsall, "Growth and Production," 232.

34. Select Committee, *Nature Conservancy*, 65.

35. *Report of The Nature Conservancy, for the year ended 30th September 1960*, 59–67. The authorship of this section of the conservancy's report is not given, but Ovington probably made a major contribution to it. He directed woodlands research at Merlewood at the time that most of the studies discussed were conducted; it emphasized the results of his own research, and many of the ideas discussed resemble those he presented at the same time and later in his own papers. Verona Conway, director of the Merlewood station during much of the 1950s, probably had Ovington's research in mind when she noted that production ecology was becoming increasingly oriented toward ecosystem energy flow and nutrient cycling (Conway, "Lines of Development," 248).

36. Ovington, "Form, Weights and Productivity," 303; Ovington, "Volatile Matter," 8; Ovington, "Comparison of Rainfall," 41–53.

37. *Report of The Nature Conservancy, for the year ended 30th September 1956*, 9.

38. Ovington, "Energy Flow," 12–20; Ovington and Heitkamp, "Accumulation of Energy," 639–646.

39. Ovington, "Quantitative Ecology," 105–106.

40. Ibid.

41. See, e.g., Carlisle, Brown, and White, "Litter Fall," and Carlisle, Brown, and White, "Nutrient Content." On the IBP see Duvigneaud, *Productivity*.

42. Golley, *Ecosystem Ecology*. Cragg, of the University of Durham's department of zoology, succeeded Ovington as director of the Merlewood Research Station.

43. Golley, *Ecosystem Ecology*, 2.

44. Description of the reserve is in Hywel-Davies, Thom, and Bennett, *Britain's Nature Reserves*, 102–103. In 1915 the SPNR had nominated Roudsea Wood as a reserve, as had the BES in 1943, the NRIC in 1945, and the Wildlife Committee in 1947 (Sheail, *Nature in Trust*, 219).

45. Nature Conservancy, *Report, 1954*, 8. Through the cooperation of the owner these surveys began even before Roudsea Wood was declared a National Nature Reserve on September 26, 1955.

46. Rackham, *History of the Countryside*, 65–67.

47. *Report of The Nature Conservancy, for the year ended 30th September 1958*, 10.

48. Nature Conservancy, *Report, 1956*, 28–29; Nature Conservancy, *Report, 1960*, 66.

49. Ibid., 66–67.

50. *Report of The Nature Conservancy, for the year ended 30th September 1957*, 2.

51. Select Committee, *Nature Conservancy*, 25.

52. Green, *Countryside Conservation*, 50.

53. Ovington, *Woodlands,* 114–142.

54. Ovington, "Form, Weights and Productivity," 302; Ovington, "Volatile Matter," 8; Ovington, *Woodlands,* 82–113.

55. Pearsall, "Habitat, Vegetation and Man," 367–368.

56. Pearsall, "Growth and Production," 235–237.

57. Champion, "Future of Our New Forests," 319.

58. See, e.g., Ovington, "Soils pH," 33.

59. Laurie, "Productivity of Forests," 77–78.

60. In 1949 H. G. Champion, chairman of the forestry subsection of the BAAS, noted the need for research on each of these concerns. Champion, "Future of Our New Forests," 320–321, 324.

61. Laurie, "Productivity of Forests"; Advisory Council on Scientific Policy, *Annual Report 1962–63,* 36.

62. An account of PEP and Nicholson's contribution to it is found in Pinder, *Fifty Years.*

63. E. Max Nicholson, letter to author, January 26, 1990. This separation was reflected in his study of government, *The System: The Misgovernment of Modern Britain.* His study was published after he had spent almost fifteen years with the conservancy, but he nevertheless did not mention it. It was also reflected in the mutually exclusive communities of the conservancy and PEP: besides Nicholson, apparently only Julian Huxley was involved in both organizations.

64. WLC/39: "Rough notes for section on report on legislation arising out of the discussion at the committee's meeting on 17th April" (E. M. Nicholson, May 8, 1946, PRO, HLG 93–48).

65. Nature Conservancy, *Report, 1954,* 5.

66. Interdepartmental Committee on the Nature Conservancy, "First Meeting Minutes," January 23, 1956. "Note by the chairman" [E. W. Playfair of the Treasury], February 29, 1956, PRO, HLG 92–180.

67. Kimball, "Nature Conservancy," 1706.

68. Minutes, Scientific Policy Committee of the Conservancy, January 28, 1954, PRO, MAF 131–25.

69. Nicholson, "Ecological Research and Farming," 439; Nature Conservancy, *Report, 1952,* 4; Nature Conservancy, *Report, 1955,* 1; "The Nature Conservancy: Agricultural Liaison—England and Wales," February 11, 1955, PRO, MAF 131–25.

70. Yapp, "Introduction," ix–xii.

71. Ecologists began a study of red grouse on the Scottish moors in 1956; their study was taken over by the conservancy in 1960. As the grouse was a popular game-bird, landowners and hunters took a great interest in this study. However, the ecologists began with several years of study devoted to building a basic understanding of red grouse demography. Only then did they feel able to give practical advice on increasing grouse populations. "We were able to work on our main aim," one ecologist recalled, "with theirs as something we often thought about for

later work when our understanding made it opportune" (Taylor, "Objective and Experiment," 31–32).

72. Nature Conservancy, *Report, 1960*, 60–61. See also Ovington, "Biological Considerations," 83, 89.

73. Nature Conservancy, *Report, 1957*, 59.

74. *Report of The Nature Conservancy, for the year ended 30th September 1961*, 5.

75. Lowe, "Values and Institutions," 343–344.

76. Moore was the conservancy's regional officer for Southwest England from 1953 to 1960, and he directed the toxic chemicals and wildlife section at Monks Wood Research Station from 1960 to 1974 (see Moore, *Bird of Time*).

77. Sheail, "Applied Ecology," 75.

CHAPTER 4. ECOSYSTEMS AND THE ATOM AT OAK RIDGE NATIONAL LABORATORY

1. Huxley, *Memories*, 174.

2. *Physics Today*, July 1950, 35, quoted in Kevles, *The Physicists*, 367.

3. Stannard, *Radioactivity and Health*, 751.

4. Balogh, *Chain Reaction*, 99–119; Forman, "Behind Quantum Electronics," 149–229.

5. Political scientists have devoted extensive study to the structure and function of iron triangles. Balogh, *Chain Reaction*, 62–64, provides an entry into this literature.

6. Hacker, *Dragon's Tail*.

7. Morgan, "Foreword," 1–2.

8. Krumholz, "Summary of Findings."

9. Seidel, "Home for Big Science," 156–157.

10. Stannard, *Radioactivity and Health*, 751.

11. Auerbach, Stanley I., interview with author, Oak Ridge, August 22, 1989.

12. Woodwell, "BRAVO Plus 25 Years," 61–64.

13. One administrator's comment about the AEC's need for an ecological research program epitomized this lack of concern: "The argument that we do not know the consequences of radiation damage to the environment is not a valid argument for support of such a program" (Memorandum dated September 16, 1954, quoted in Whicker and Schultz, *Radioecology*, 5). See also comments of Abel Wolman, an environmental engineer, concerning the lack of concern among AEC physicists and chemists regarding the environment, in Balogh, *Chain Reaction*, 153.

14. Oak Ridge National Laboratory, *Health Physics Division Progress Report, Period Ending July 31, 1956*, 18–20; Auerbach, "Soil Ecosystem," 527.

15. Pritchard and Joseph, "Disposal of Radioactive Wastes," 29.

16. Morgan and Auerbach, "Melton Valley," 1.

17. Auerbach, Stanley I., interview with author, Oak Ridge, August 22, 1989; United

States Atomic Energy Commission, *19th Semi-Annual Report, Period Ending January 31, 1956,* 70.

18. Oak Ridge National Laboratory, *Health Physics Division Progress Report, Period Ending July 31, 1955,* 7–8; ORNL, *Health Physics Report, 1956,* 5–9.

19. Oak Ridge National Laboratory, *Health Physics Division Annual Progress Report, Period Ending July 31, 1959,* 25–27; Oak Ridge National Laboratory, *Health Physics Division Annual Progress Report, Period Ending July 31, 1961,* 87–90; Auerbach and Crossley, "Cesium-137 Uptake," 494–499.

20. Oak Ridge National Laboratory, *Health Physics Division Progress Report, Period Ending July 31, 1957,* 27–37; *Health Physics Division Annual Progress Report, Period Ending July 31, 1958,* 48; *Health Physics Division Annual Progress Report, Period Ending July 31, 1960,* 147; Cowser and Parker, "Soil Disposal," 152–163; Morton and Struxness, "Ground Disposal," 156–163.

21. ORNL, *Health Physics Report, 1957,* 10; ORNL, *Health Physics Report, 1958,* 38.

22. ORNL, *Health Physics Report, 1960,* 186.

23. Not until the late 1960s did the AEC again support research into ecological aspects of waste disposal. See Del Sesto, *Civilian Nuclear Power,* 98–99, and Lipschutz, *Radioactive Waste,* 168. For a thorough discussion of waste disposal and the AEC see Mazuzan and Walker, *Controlling the Atom,* chap. 12.

24. Odum had been associated with the AEC since 1951; see E. P. Odum, "Total Environment," 350–353.

25. Quoted in Auerbach, "Evolution," 1.

26. Auerbach, "Long-Term Role," 2.

27. Auerbach, Stanley I., interview with author, Oak Ridge, August 21, 1989. The role of this committee, he explained, was to illustrate "what ecology was to a laboratory management which admittedly was interested in what ecology is, but really didn't quite understand it" (Auerbach and Millemann, *Dedication,* 31).

28. Lindeman, "Trophic-Dynamic Aspect," 319–418. See also Cook, "Raymond Lindeman," 22–26; Hagen, *Entangled Bank,* 87–99.

29. Elton, *Animal Ecology.*

30. Lindeman, "Trophic-Dynamic Aspect," 399–400.

31. Hutchinson and Wollack, "Connecticut Lake Sediments," 483–517.

32. Hutchinson, "Bio-Ecology," 267–268.

33. Hutchinson, "Circular Causal Systems in Ecology," 221–246.

34. Taylor, "Technocratic Optimism," 215–220.

35. H. T. Odum, "World Strontium Cycle," 407–411; H. T. Odum, "Productivity of Silver Springs," 55–112.

36. H. T. Odum and R. C. Pinkerton, "Time's Speed Regulator," 331–343; see Taylor, "Technocratic Optimism," 226–244.

37. Teal, "Salt Marsh Ecosystem," 614–624; Margalef, "Information Theory in Ecology," 36–71; Patten, "Cybernetics of the Ecosystem," 221–231.

38. Besides Clements, other influential organicist ecologists included members of

the "Chicago school of ecology." Their perspective is described in Allee et al., *Principles of Animal Ecology;* see also Mitman, *State of Nature*.

39. Hollaender, "Biology Division," 10–14.

40. ORNL, *Health Physics Report, 1961,* 105–106.

41. Auerbach and Schultz, *Onsite Ecological Research,* 1.

42. Auerbach, "Space Program," 21–22.

43. Crossley, "Consumption of Vegetation by Insects," 427–430; Crossley, "Use of Radioactive Tracers," 43–53; Crossley, "Movement and Accumulation," 103; Crossley and Howden, "Insect-Vegetation Relationships," 302.

44. Crossley and Reichle, "Analysis of Transient Behavior," 341–343; Reichle, "Radioisotope Turnover," 351–366; Reichle, "Forest Floor Arthropods," 538–542; Reichle, Dunaway, and Nelson, "Turnover and Concentration," 43–45.

45. ORNL, *Health Physics Report, 1960,* 156; Kaye and Dunaway, "Bioaccumulation," 205–217; Kaye and Dunaway, "Estimation of Dose Rate," 107–111.

46. Olson, "Rates of Succession," 166, 168.

47. ORNL, *Health Physics Report, 1960,* vi, 174–175; ORNL, *Health Physics Report, 1961,* 105; Oak Ridge National Laboratory, *Health Physics Division Annual Progress Report, Period Ending July 31, 1962,* 56; Auerbach and Olson, "Biological and Environmental Behavior," 515.

48. Charles Dunham, director of the AEC's Division of Biology and Medicine, explained that at the time the commission had neglected study of other radionuclides found in fallout because "we were too busy chasing Strontium 90" (Metzger, *Atomic Establishment,* 108).

49. ORNL, *Health Physics Report, 1960,* 167–168.

50. Hagen, *Entangled Bank,* 112–115.

51. Auerbach, Stanley I., interview with author, Oak Ridge, August 22, 1989; Crossley, "Use of Radioactive Tracers," 52.

52. ORNL, *Health Physics Report, 1959,* 41–45; ORNL, *Health Physics Report, 1960,* 168–169; Neel and Olson, "Use of Analog Computers."

53. Olson, "Equations for Cesium Transfer," 1385–1392.

54. ORNL, *Health Physics Report, 1959,* 41. See also ORNL, *Health Physics Report, 1960,* 167, and Oak Ridge National Laboratory, *Health Physics Division Annual Progress Report, Period Ending July 31, 1965,* 92.

55. ORNL, *Health Physics Report, 1959,* 45; Olson, "Energy Storage," 322–331; ORNL, *Health Physics Report, 1965,* 92.

56. E. P. Odum, "Panel Discussion," 643–645; E. P. Odum, "Feedback," 1257–1262.

57. E. P. Odum, "Feedback."

58. See, e.g., Patten, "Cybernetics of the Ecosystem," 221–231, and Patten, "Competitive Exclusion," 1599–1601. The significance of cybernetics and other areas of research concerned with the behavior of systems to ecology is discussed in Hagen, *Entangled Bank,* 130–133.

59. Oak Ridge National Laboratory, *Health Physics Division Annual Progress Re-*

port, *Period Ending July 31, 1964*, 104–108; ORNL, *Health Physics Report, 1965*, 89–92; Patten and Witkamp, "Systems Analysis," 813–824.

60. Auerbach, Stanley I., interview with author, Oak Ridge, August 22, 1989; Van Dyne, "State of the Art," 88.

61. Auerbach recalled this, in comments introducing Van Dyne's paper, in "State of the Art," 81.

62. ORNL, *Health Physics Report, 1965*, 97, 99; Oak Ridge National Laboratory, *Health Physics Division Annual Progress Report, Period Ending July 31, 1966*, 109–119; Van Dyne, "Operations Research Techniques," 1511–1519. Van Dyne discussed many of his ideas on systems research in "Ecosystems, Systems Ecology, and Systems Ecologists."

63. ORNL, *Health Physics Report, 1964*, 105–106.

64. Patten, "Systems Ecology," 593. See also Patten, "Systems Approach."

65. Patten, "Community Organization"; Auerbach, Stanley I., interview with author, Oak Ridge, August 22, 1989.

66. Nelson and Scott, "Role of Detritus," 396–413.

67. Curlin and Nelson, "Walker Branch Watershed," 5–11.

68. Balogh, *Chain Reaction*, 188–189; Weinberg, *Reflections*, 132. In contrast, less than 1 percent of research at the Brookhaven and Argonne laboratories was funded by outside agencies. See Steinhart and Cherniack, *Problem Focused Education*, 14, and Balogh, *Chain Reaction*, 298.

69. Auerbach, "Space Program," 22; Auerbach, "Long-Term Role," 1.

70. Balogh, *Chain Reaction*, 297–298.

71. Kaye and Ball, "Systems Analysis," 731–739. See also Booth and Kaye, "Radioactivity Transfer," and Booth, Kaye, and Rohwer, "Predicting Dose to Man," 877–893.

72. Oak Ridge National Laboratory, *Health Physics Division Annual Progress Report, Period Ending July 31, 1969*, 98–110.

73. Auerbach, Stanley I., interview with author, Oak Ridge, August 21, 1989. To a large extent, of course, such a view reflected the low status of ecology within the national scientific community during the 1950s and early 1960s; see Kwa, "Representations," 416–417.

74. Auerbach, "Soil Ecosystem," 523.

75. Auerbach, Olson, and Waller, "Landscape Investigations," 761.

76. Weinberg, "Large-Scale Science," 161–164.

77. ORNL, *Health Physics Report, 1961*, 124. On the use by physiologists and health physicists of these techniques see Robertson, "Biological Systems," 133–154.

78. Olson, "Analog Computer Models," 121.

79. Kwa, "Representations," 422–426.

80. Commenting on computer simulations of reactors in 1972, Weinberg stated, "As an old-timer who grew up in this business before the computing machine dominated it so completely, I have a basic distrust of very elaborate calculations of complex situations, especially where the calculations have not been checked by

full-scale experiments" (Weinberg to James Schlesinger, chairman, AEC, February 9, 1972, in Primack and von Hippel, *Advice and Dissent*, 221.

81. Ovington and Lawrence, "Strontium 90," 175–181.

82. National Academy of Sciences, *Federal Support*, 28.

83. Anderson, "Production Facilities," 83–84.

84. Walker, "Nuclear Power and the Environment," 968, 975–979.

85. Balogh, *Chain Reaction*.

86. Mazuzan and Walker, *Controlling the Atom*, 250–251.

87. Ibid., 345–372. This was in spite of concerns about the impact of the larger quantities of waste expected with the development of nuclear energy, as expressed by Eugene Odum, among others; see E. P. Odum, *Fundamentals of Ecology*, 2nd ed., 481.

88. Contemporary discussions of this assumption include Greenberg, *Politics of Pure Science;* Brooks, *Government of Science;* Reagan, *Science and the Federal Patron*.

89. Forman, "Behind Quantum Electronics."

CHAPTER 5. OAK RIDGE ECOSYSTEM RESEARCH AND IMPACT ASSESSMENT

1. Walker, "Atomic Energy Commission," 57–78; Lindop and Rotblat, "Radiation Pollution of the Environment," 17; Dawson, *Nuclear Power*, 206.

2. Tamplin, "Issues in the Radiation Controversy," 25–27; Gofman and Tamplin, *Poisoned Power;* Worster, *Nature's Economy*, 339.

3. Keating, "Technology Assessment," 59–64; Gillette, "Reactor Emissions," 1215; Boffey, "Radioactive Pollution," 1043–1046; Balogh, *Chain Reaction*, 266.

4. Brungs, "Heated Water,"; Cairns, "We're in Hot Water," 10; Clark, "Thermal Pollution and Aquatic Life," 220.

5. Nelkin, *Nuclear Power and Its Critics;* Walker, "Nuclear Power and the Environment," 964–992. Nelkin also noted that a major obstacle to resolving the Cayuga Lake controversy was the difficulty experienced by both sides in obtaining data from which to draw definitive conclusions. This fact would be significant in later controversies, as is discussed later in this chapter.

6. In 1969 the AEC fended off a legal challenge to this narrow interpretation. AEC, "Jurisdiction of the AEC and Other Federal and State Agencies with Respect to Nuclear Power Facilities," 2, quoted in Keating, "Technology Assessment," 55–56; Bronstein, "Calvert Cliffs," 704.

7. AEC, *Nuclear Power and the Environment*, 29, quoted in Keating, "Technology Assessment," 41–42.

8. Weinberg, "Nuclear Energy and the Environment," 71; Weinberg, "Social Institutions and Nuclear Energy," 29.

9. Del Sesto, *Civilian Nuclear Power*, 91; Walker, "Nuclear Power and the Environment," 964–968.

10. Holifield, "Remarks," 927, quoted in Keating, "Technology Assessment," 66. See

also Dawson, *Nuclear Power*, 241. On the iron triangle see Balogh, *Chain Reaction*.

11. Balogh, *Chain Reaction*, 262.

12. Moravcsik, "Reflections on National Laboratories," 16.

13. Quoted in Rose, "New Laboratories for Old," 146.

14. Walsh, "Federal Laboratories," 442–444.

15. Quote from Gillette, "Roots of Dissent," 773. On the views of laboratory scientists see Gillette, "Years of Delay," 870.

16. Anderson, "Production Facilities," 78; Dawson, *Nuclear Power*, 26–27.

17. Reichle and Crossley, "Heterotrophic Productivity," 563–587.

18. Battelle, *Evaluation*, II-140.

19. Auerbach, "Biome Research Proposal," 7.

20. Oak Ridge National Laboratory, *Ecological Sciences Division, Annual Progress Report, For Period Ending September 30, 1971*, 105–108.

21. See, e.g., Sollins, "Organic Matter Budget," and Goldstein and Harris, "SERENDIPITY." Quote from ORNL, *Ecological Sciences Division, Report, 1971*, viii.

22. O'Neill, "Modeling," 56–59.

23. Burgess and O'Neill, "Progress Report," 327–331.

24. Shugart et al., "TEEM," 261. See also O'Neill et al., "Terrestrial Ecosystem Energy Model." O'Neill, one of the leading modelers at Oak Ridge, once noted that "frustrating efforts have convinced virtually everyone that assembling complex process models does not necessarily result in a realistic ecosystem model" (O'Neill, "Modeling," 64).

25. Shugart et al., "TEEM," 261.

26. Reichle et al., "Modeling Forest Ecosystems," iii; Auerbach, "Biome Research Proposal."

27. Auerbach, "Structure and Function," 39.

28. Oak Ridge National Laboratory, *Environmental Sciences Division, Annual Progress Report, For Period Ending September 30, 1973*, vii.

29. O'Neill, "Tracer Kinetics," 693; Reichle and Auerbach, "Analysis of Ecosystems," 266–273; Burgess and Auerbach, "Objectives," 15; Auerbach, "Deciduous Forest Biome Program," 680.

30. Alvin M. Weinberg to John R. Totter, June 6, 1968, Auerbach personal files.

31. Hays, *Beauty, Health, and Permanence*.

32. Averch, *Strategic Analysis;* Caldwell, "Environment," 132–139; Hurst, *Social Order*.

33. Lilienfeld, *Systems Theory;* Kwa, in "Representations," argues that describing the American IBP program to Congress as a systems approach helped ensure its funding. U.S. Congress, House of Representatives, *Environmental Pollution*, 4.

34. Reichle and Auerbach, "Analysis of Ecosystems," 261, 271.

35. O'Neill and Burke, "Simple Systems Model"; Loucks, "Environmental Court Actions," 419–473; Hett, "Land-Use Changes"; Burgess and Kern, "Progress Report," 188; ORNL, *Environmental Sciences Division, Report, 1973*, 98–99.

36. See, e.g., the attempt of the Environmental Systems Group at the University of California, Davis, to model the state of California in Foin, "Human Society," 475–531. Among the most prominent proponents of this view were Eugene and Howard Odum. E. Odum, for example, suggested in 1972 that "environmental science is now being called upon to help determine a realistic level of human population density and rate of use of resources and power that are optimum in terms of the quality of human life, in order that 'societal feedback' can be applied before there are serious overshoots" (Odum, "Ecosystem Theory," 13; see also Taylor, "Technocratic Optimism").

37. Entry into this debate is provided by Woodwell and Smith, *Diversity and Stability;* and May, *Stability and Complexity*.

38. Reichle, "Advances in Ecosystem Analysis," 257–264; O'Neill et al., "Theoretical Basis," 28–40.

39. O'Neill et al., "Monitoring Terrestrial Ecosystems," 271–277.

40. Olson, "Carbon Cycles," 227.

41. Harris and Reichle, "Forest Productivity," 57.

42. A survey conducted shortly after the conclusion of the IBP found that researchers of the EDFB and two other biomes did not feel pressed to justify their projects to biome management in terms of their contribution to the entire program (Battelle, *Evaluation*, II-112).

43. Peter Galison has made a similar observation of the large collaborations characteristic of experimental physics; see Galison, *How Experiments End*, 263–278.

44. Kwa, "Representations"; McIntosh, *Background*, 213–234; Golley, *History*, 120–135; Kwa, "Modeling the Grasslands."

45. Auerbach, "Response to Stress," 33.

46. Oak Ridge National Laboratory, *Environmental Sciences Division, Annual Progress Report, Period Ending September 30, 1977;* Oak Ridge National Laboratory, *Environmental Sciences Division, Annual Progress Report, Period Ending September 30, 1978.*

47. Keating, "Technology Assessment," 37–74; Bronstein, "Calvert Cliffs," 689–721.

48. ORNL, *Environmental Sciences Division, Report, 1973*, 19.

49. Auerbach, "Ecology, Ecologists and the E. S. A.," 4.

50. Oak Ridge National Laboratory, *Environmental Sciences Division, Annual Progress Report, Period Ending September 30, 1972*, vii. In 1972 Auerbach noted the "reluctance and disinterest shown by some young ecologists" toward impact assessment work; see Auerbach, "Ecology, Ecologists and the E.S.A.," 4.

51. Hooper, "Applications of Ecology," 45–51.

52. Eberhardt, "Quantitative Ecology," 29.

53. Christensen, Van Winkle, and Mattice, "Defining and Determining," 192.

54. Bronstein, "Calvert Cliffs," 720; Gelpe and Tarlock, "Uses of Scientific Information," 371–427.

55. Auerbach, "Current Perceptions," 166; Van Winkle, Christensen, and Mattice, "Two Roles," 254.

56. Greenwood, *Knowledge and Discretion,* 252.

57. Smith and Wynne, *Expert Evidence;* see also Nelkin, *Controversy.*

58. Lindop and Rotblat, "Radiation Pollution of the Environment."

59. Quote from Auerbach et al., "Understanding," 10; O'Neill, "Pathway Analysis"; Auerbach et al., "Ecological Considerations," 811, 813; Booth, Kaye, and Rohwer, "Systems Analysis Methodology," 877–893.

60. Reichle, Dunaway, and Nelson, "Turnover and Concentration," 52.

61. Oak Ridge National Laboratory, *Environmental Sciences Division, Annual Progress Report, Period Ending September 30, 1976,* 2.

62. Ibid., 2.

63. Talbot, *Power Along the Hudson.*

64. Van Winkle, "Application of Computers," 90.

65. Christensen et al., "Defining and Determining," 205; Barnthouse et al., "Introduction to the Monograph," 1–8.

66. Auerbach, "Current Perceptions," 164–165.

67. Christensen et al., "Defining and Determining," 202; Van Winkle et al., "Two Roles," 248.

68. Auerbach, "Current Perceptions," 166.

69. Hall, "Decision Making Process," 345–364.

70. Barnthouse et al., "Population Biology," 16.

71. Christensen et al., "Defining and Determining," 208–210.

72. Ibid., 210.

73. Ibid., 208.

74. Englert and Boreman, "Historical Review," 143–151; Van Winkle et al., "Impingement Losses," 29–33.

75. Christensen and Englert, "Historical Development," 133–142.

76. Smolowe, "Storm King Plant"; Barnthouse et al., "Settlement Agreement," 267–273.

77. Oak Ridge National Laboratory, *Environmental Sciences Division, Annual Progress Reports,* 1972–1979; Auerbach, "Space Program," 21–25.

78. ORNL, *Ecological Sciences Division, Report, 1971,* 134.

79. This is only a brief summary of a large literature on the administration of American environmental regulations and policy. See Greenwood, *Knowledge and Discretion;* Jasanoff, "Problem of Rationality," 151–183; Mann, *Environmental Policy Implementation;* Marcus, *Promise and Performance;* and Rosenbaum, *Politics of Environmental Concern.*

80. Dickson, *New Politics of Science,* 261–306.

81. On this "crisis of authority" and its relation to environmental controversy see Nelkin, "Political Conflict," ix–xxv.

82. This was not absolute. In some instances ecosystem research was applied to such problems. For example, lake ecosystem models developed with the assistance of Oak Ridge ecologists were eventually applied to environmental management.

83. See, e.g., White, "Environment," 186; Caldwell, *Man and His Environment,* 85.

84. Bunnell, "Theological Ecology," 170.

85. Auerbach, "Space Program," 23–24.

86. This building was completed in 1978.

87. It evidently has considerable public relations value: in 1989 80 percent of the exhibits at the laboratory's visitors' center described environmental research, which constitutes only about 4 percent of the laboratory's activities.

88. Nelkin, "Professional Responsibility," 75–95.

89. Auerbach, Stanley I., interview with author, Oak Ridge, August 22, 1989.

CHAPTER 6. FOREST EXPERIMENT AND PRACTICE AT HUBBARD BROOK

1. Bormann, "Loblolly Pine and Sweetgum," 339–358.

2. Bormann noted the influence of two texts on forest watersheds—Kittredge, *Forest Influences*, and Colman, *Vegetation and Watershed Management*—in "Lessons From Hubbard Brook," 1.

3. Bormann, F. Herbert, interview with author, July 8, 1989; Bormann, "Primary Leaf," 189–192; Bormann, "Percentage Light Readings," 473–476; Bormann, "Moisture Transfer," 48–55.

4. Bormann and Graham, "Natural Root Grafting," 677–691.

5. Pearsall, *Mountains and Moorlands,* 113–114, 188–190.

6. Bormann, F. Herbert, interview with author, July 8, 1989.

7. F. Herbert Bormann to Robert S. Pierce, November 7, 1960. Bormann had by this date worked out the basic concept of what would become the Hubbard Brook Ecosystem Study. An excerpt of this letter follows:

The other day while discussing the problem of mineral cycling through ecosystems, the thought came to me that your installation at Hubbard Brook represents a veritable research gold mine in regard to fundamental studies on mineral cycling.

One of your small watersheds with a weir at the outlet represents a perfect area for controlled research. If one were to select one or several minerals, such as $K+$, it would be possible, by taking weekly water samples and analyzing them, to determine quantitatively the amount of $K+$ leaving the system. Since the watershed is theoretically tight and all water falling on the shed appears at the weir, the quantitative figure would represent total loss of $K+$ from the watershed.

Some minerals may be added by rain or snowfall, therefore both rain and snow would have to be analyzed for the mineral(s) in question. These analyses multiplied by the amount of rain or snow would give the total amount of the mineral(s) added to the system.

By subtracting the total amount added from the total amount lost, it would be possible to estimate the steady-state losses from the system. Theoretically the only place these minerals could come from is the underlying parent material and bedrock. Thus, the loss represents the rate at which the bedrock is wasting away in terms of the mineral(s) under consideration. By knowing the chemical composition of the bedrock, it would be possible to determine the rate at which it was breaking down.

8. Pierce, Robert S., interview with author, July 7, 1989.
9. Quoted in Baldwin and Brooks, *Forests and Floods*, 2.
10. Marsh, *Artificial Propagation of Fish*, 14.
11. Robinson, *Forest Service*, 246; Schiff, *Fire and Water*, 116–163; Steen, *U.S. Forest Service*, 131–141.
12. Steidtman, "Geochemical Cycle of Iodine," HBES Files. Bormann suggested the topic, and Steidtman was supervised by Robert Reynolds, a geology professor.
13. Hasler, "Experimental Limnology," 36.
14. Ibid., 36.
15. Likens and Hasler, "Movements of Radiosodium," 48–56; Ragotzkie and Likens, "Heat Balance," 412–425; Likens, Gene E., interview with author, July 11, 1989.
16. According to Likens and Bormann, the reviews of their proposal were not favorable, and they believed that Bormann's friendship with the NSF project director, George Sprugel, was in part responsible for its approval (Bormann, F. Herbert, interview with author, July 8, 1989; Likens, Gene E., interview with author, July 11, 1989).
17. Bormann completed his studies of root grafting; Likens wrote up his studies of the limnology of Alaskan and Antarctic lakes (Likens, "Unusual Distribution of Algae," 213–217; Ragotzkie and Likens, "Heat Balance").
18. Bormann and Likens, "Nutrient Cycling," 429.
19. Bormann, Likens, and Johnson, "Hydrologic-Mineral Cycle Interaction," 5–6.
20. Bormann and Likens, "Nutrient Cycling," 425, fig. 1.
21. Ovington, "Quantitative Ecology," 176.
22. Bormann and Likens, "Nutrient Cycling," 424–425.
23. Likens et al., "Small Forested Ecosystem," 772–773.
24. Bormann and Likens, "Nutrient Cycling," 426.
25. Ibid., 426.
26. Ibid., 427; Johnson et al., "Chemical Weathering," 531–545.
27. Likens, Gene E., interview with author, July 11, 1989.
28. See, e.g., Bormann and Graham, "Natural Root Grafting," 678, and Bormann, "Percentage Light Readings," 473.
29. Beckel, "Breaking New Waters," 31.
30. Bormann, F. Herbert, interview with author, July 8, 1989; Likens, Gene E., interview with author, July 11, 1989.
31. Likens et al., "Effects of Forest Cutting," 23–47.
32. Hoover, "Removal of Forest Vegetation," 969–975; Hornbeck, Pierce, and Federer, "Streamflow Changes," 1124–1132.
33. Likens et al., "Effects of Forest Cutting," 31.

34. Colman, *Vegetation and Watershed Management*, cited in Bormann, Likens, and Eaton, "Biotic Regulation," 600–610.

35. Bormann and Likens, *Pattern and Process*, 59.

36. F. Herbert Bormann to Gene E. Likens, December 2, 1968; Robert S. Pierce to Gene E. Likens February 14, 1969, HBES Files.

37. Likens et al., "Effects of Forest Cutting."

38. Johnson et al., "Working Model," 1353–1363; Bormann et al., "Biotic Regulation"; Likens et al., "Effects of Forest Cutting"; Johnson et al., "Chemical Weathering."

39. Likens et al., "Effects of Forest Cutting," 39.

40. Smith, Bormann, and Likens, "Chemoautotrophic Nitrifiers," 471–473; Likens, Bormann, and Johnson, "Nitrification," 1205–1206.

41. Bormann, "Lessons from Hubbard Brook," 4.

42. Likens, Gene E., interview with author, July 11, 1989.

43. Bormann et al., "Nutrient Loss," 882–884; Likens et al., "Nitrification"; Dominski, "Accelerated Nitrate Production"; Gosz, Likens, and Bormann, "Nutrient Content," 769–784; Gosz, Likens, and Bormann, "Nutrient Release," 173–191; Gosz, Likens, and Bormann, "Organic Matter," 305–320.

44. Robert S. Pierce to Gene E. Likens, December 22, 1966, HBES Files.

45. Robert S. Pierce to Gene E. Likens, May 10, 1969, and K. G. Reinhart to Robert S. Pierce, April 3, 1969, HBES Files.

46. Robert S. Pierce to Gene E. Likens, May 10, 1969, and accompanying title page of draft paper, HBES Files.

47. It had been decided to discuss the hydrologic aspects of the experiment in a separate paper.

48. Likens and Bormann, "Effects of Forest Clearing," 330–335.

49. Robert S. Pierce to Gene E. Likens, February 14, 1969, HBES Files.

50. Likens, et al., "Effects of Forest Cutting," 43–45.

51. Bormann and Likens, "Watershed-Ecosystem Concept," 50.

52. "Tree Cutting and Nutrient Loss," 272.

53. U.S. Senate Subcommittee on Public Lands, *National Timberlands*, 157–173.

54. Earl L. Stone to Robert S. Pierce, September 27, 1972, HBES Files.

55. Congressman Donald Fraser to Gene E. Likens, June 9, 1972; George W. Bengtson to F. Herbert Bormann, May 7, 1971; Cooper to Bormann, April 21, 1971, HBES Files; quote from Pierce, Martin, and Likens, "Clearcutting and Nutrient Loss," 2.

56. Bormann, F. Herbert, interview with author, July 8, 1989.

57. Charles F. Cooper to F. Herbert Bormann and Gene E. Likens, August 5, 1971, HBES Files.

58. F. Herbert Bormann and Gene E. Likens to Charles F. Cooper, August 10, 1971, HBES Files.

59. Likens, Gene E., interview with author, July 11, 1989.

60. See, e.g., Federer, "Soil-Plant-Atmosphere Model," 555–562; Hornbeck, Likens, and Eaton, "Seasonal Patterns," 355–365; Martin, "Precipitation and Streamwater Chemistry," 36–42.

61. Hornbeck and Likens, "Composition of Snow," 139–151.

62. Hays, *Beauty, Health, and Permanence,* 394–397.

63. Likens had taken advantage of a small NSF program to support undergraduate research, but the results didn't seem to him worth the time taken to supervise the inexperienced students in the field. Likens, "Final Report," 5.

64. Student figures compiled from Likens, *Publications;* Likens, *Mirror Lake.* As an evaluation study of the IBP noted (Battelle, *Evaluation,* I-39-I-41), the cost of the Hubbard Brook study was much less per paper published than the IBP programs. This difference was attributed to the gradual evolution and greater maturity of the Hubbard Brook study. The participation of students also helps explain its economic efficiency, however.

65. Bormann and Likens, *Pattern and Process,* 144–145, reproduces pictures showing the annual growth of the vegetation; quote from Bormann and Likens, "Hydrologic-Nutrient Cycle Interaction," 1976 proposal, 138. The recovery was described in Likens et al., "Recovery," 492–496.

66. Marks and Bormann, "Revegetation," 915; see also Marks, "Role of Pin Cherry," 73–88.

67. Anonymous review accompanying Arthur W. Cooper to F. Herbert Bormann, December 7, 1972, HBES Files.

68. Bormann et al., "Export of Nutrients," 255–277.

69. Bormann et al., "Composition and Dynamics," 373–388; Siccama, Bormann, and Likens, "Productivity, Nutrients and Phytosociology," 389–402.

70. Bormann, F. Herbert, interview with author, July 8, 1989.

71. Whittaker et al., "Forest Biomass and Production," 233–254; Likens et al., *Biogeochemistry,* 90.

72. Whittaker et al., "Forest Biomass and Production," 253.

73. F. Herbert Bormann to Robert H. Whittaker, August 24, 1972, HBES Files.

74. Forcier and Bormann, "Intraspecific Energy Flow Model," 92–93; Bormann, F. Herbert, letter to author, May 28, 1992.

75. Siccama was initially involved in this effort but did not continue, preferring to contribute to studies of forest growth that did not involve computers.

76. Botkin, Janak, and Wallis, "Computer Model," 849–872.

77. Whittaker et al., "Forest Biomass and Production," 250; Botkin, Janak, and Wallis, "Computer Model," 849–872.

78. Botkin, "Anthropogenic Production," 325–331.

79. Bormann, "Lessons from Hubbard Brook," 3; Bormann and Likens, "Hydrologic-Nutrient Cycle Interaction," 1976 proposal, 1.

80. Fisher and Likens, "Stream Ecosystem," 33–35; Fisher and Likens, "Bear Brook," 421–439.

81. Bormann and Likens, "Hydrologic-Mineral Cycle Interaction," 1969 proposal.

82. Holmes and Sturges, "Annual Energy Expenditure," 24–29; Sturges, Holmes, and Likens, "Role of Birds," 149–155; Holmes, Sherry, and Bennett, "Diurnal

and Individual Variability," 141–149; Holmes, Bonney, and Pacala, "Guild Structure," 512–520; Holmes, Richard, interview with author, July 6, 1989.

83. Hall et al., "Experimental Acidification," 976–989; Meyer, "Leaf Decomposition," 44–53.

84. Bormann, F. Herbert, interview with author, July 8, 1989.

85. Emphasis in original. This passage was repeated in a publication (Likens et al., *Biogeochemistry*, vi); in a proposal to the NSF (Bormann and Likens, "Hydrologic-Nutrient Cycle Interaction," 1976 proposal, 1); and in response to questions posed by the NSF review panel (Likens and Bormann, "Hubbard Brook Ecosystem Study, Response to NSF Questions," February 22, 1976, 3, HBES Files).

86. Two reviews of this expanding body of research are Gorham, Vitousek, and Reiners, "Regulation of Chemical Budgets," 53–84; and O'Sullivan, "Ecosystem-Watershed Concept," 273–281.

87. Bormann, F. Herbert, interview with author, July 8, 1989.

88. This was within a passage that Bormann and Likens quoted from Alex Watt's famous paper "Pattern and Process," 22. This paper also inspired, of course, the title of their book.

89. Bormann and Likens, *Pattern and Process*, 186.

90. Ibid., v.

91. Ibid., 6.

92. Ibid., 174. The paper they cited was Vitousek and Reiners, "Ecosystem Succession," 376–381.

93. Bormann and Likens, *Pattern and Process*, 166.

94. Hagen, *Entangled Bank*, 182–183. Of course, as we have seen at Oak Ridge, not all IBP studies adhered to this model.

95. On big science see Capshew and Rader, "Big Science." The significance of long-term phenomena has been neglected in studies of big science, perhaps because most such studies have focused on the physical sciences, especially subatomic events that last only a fraction of a second (see, e.g., several papers in Galison and Hevly, *Big Science*); on the long-term significance of Hubbard Brook research see Likens, "Priority for Ecological Research."

96. Likens et al., *Biogeochemistry*, vii.

97. Bormann and Likens, *Pattern and Process*, 221–227.

98. Bormann and Likens, "Fresh Air," 62–71; Bormann, "Air Pollution Stress," 188–194; Bormann, "Effects of Air Pollution," 338–346.

99. Bormann and Likens, *Pattern and Process*, 227.

100. Ibid., 21–23 (on Covington and Dominski conclusion), 97 (on erosion), 175 (on steady-state phase), and 216 (on studies of commercially clearcut areas).

101. Hubbard Brook ecologists emphasized this contrast, and their skepticism of models: "We have no grand computerized model where all the animate and inanimate components and processes of the dynamic ecosystem are elegantly

linked and where the details of the interactions can be spilled for by a conversation with the computer. Indeed, no such model exists for any ecosystem" (*Pattern and Process*, 3).

102. Bormann and Likens, "Hydrologic-Nutrient Cycle Interaction," 1978 proposal, 16.

103. Bormann, "Inseparable Linkage," 754–760; Bormann and Likens, "Fresh Air," 62–71; Bormann, "Air Pollution Stress," 188–194; Bormann, "Effects of Air Pollution," 338–346.

104. Likens, Gene E., interview with author, July 11, 1989; on American scientists' perception of research agencies see Palmer, King, and Shannon, "Oceanographers," 73–90.

CHAPTER 7. ECOLOGY AND THE ONTARIO FISHERIES

1. Whitaker, "Early History," 172.

2. Connery, *Governmental Problems*, 11.

3. Benson, "American Fisheries Society," 13–24; McEvoy, *Fisherman's Problem*.

4. The early history of the OFRL is described in Clemens, *Education and Fish*.

5. Bensley, "Biological Investigation," 6–7, 13–20.

6. Girard, "Commission of Conservation," 19–40; Johnstone, *Aquatic Explorers*.

7. Harkness, "Ontario Fisheries Research Laboratory."

8. By 1939 the OFRL had accumulated fifty-eight papers in its publication series. Several students trained at the laboratory, including Donald S. Rawson and William A. Kennedy, went on to continue aquatic research elsewhere, particularly in western Canada.

9. Leslie, "DDT in Ontario's Forests," 186–197.

10. Richardson, *Conservation by the People;* Shrubsole, "Water Management Agencies," 49–66.

11. There was likely contact between British and Canadian ecologists at this time; however, I did not find records of this in either the Public Record Office or the University of Toronto Archives.

12. Report of the Chairman of the Advisory Committee on Fisheries and Wildlife to the Ontario Research Council, April 17, 1947, University of Toronto Archives, File A74–0022, Box 12.

13. R. N. Johnston, Chief, Division of Research, to K. H. Doan, November 27, 1951, University of Toronto Archives, File A74–0022, Box 25.

14. Christie, "Effects of Artificial Propagation," 597–639.

15. See, e.g., Dymond and Delaporte, "Pollution of the Spanish River," and Dymond, "Advisory Committee," Harkness Library, University of Toronto Department of Zoology.

16. Fry, "South Bay Experiment," 15–25; Budd, "Tagged Whitefish," 128–134; Fraser, "Smallmouth Bass Fishery," 147–177.

17. Fry, "Angler's Arithmetic," 61.

18. See, e.g., Lawler, "Fluctuations," 1197–1227; Fry and Watt, "Yields of Year Classes," 136–143; and Christie, "Effects of Artificial Propagation."

19. Lapworth, "Effect of Fry Plantings," 547–558; Christie, "Effects of Artificial Propagation"; Dymond, "Management of Great Lakes Fisheries," 384–392.

20. Dymond, "Fisheries of the Great Lakes," 87.

21. Whillans, "F. E. J. Fry's Field Studies," 574–584; Elton, *Animal Ecology*, 41–49.

22. Laboratory for Experimental Limnology, "Report of the Laboratory for Experimental Limnology for the Years 1952 and 1953," and "The Laboratory for Experimental Limnology, 1954–1959." Fry set out the conceptual basis for much of this research in Fry, "Effects of the Environment," 5–62.

23. Langford, "Fertilization of Lakes," 133–144.

24. Harkness and Fry, "Game Fish Management," 398–404; Langford, "Fisheries Research," 25–32; Langford, "Opeongo Laboratory," 25–28.

25. For example, after nutrients were added to lakes, the impact on phytoplankton and zooplankton, as well as fish, was examined.

26. Dymond, "Fisheries of the Great Lakes," 90.

27. Quoted in Dymond, "Fisheries of the Great Lakes," 64.

28. Ontario Department of Lands and Forests, *Annual Report*, 1962, 257.

29. Fry, F. E. J., to "Prof. D." (probably J. R. Dymond) January 25, 1956, File A74–0022, Box 25, University of Toronto Archives. See, e.g., Kenneth Watt's sophisticated analyses of fisheries populations, pursued during the 1950s at the Maple Laboratory: Watt, "Choice and Solution," 613–645, and Watt, "Studies on Population Productivity," 367–392.

30. Dymond, "Problems of Fisheries."

31. Doern, *Science and Politics in Canada*.

32. Fry, "1944 Year Class," 178–192; Coble, "White Sucker Population," 2117–2136.

33. Ryder, "Fisheries Inventories in Ontario"; Harvey and Geen, "Introduction to Limnology," 501–504; Lambert and Pross, *Renewing Nature's Wealth;* Ryder, "Fundamental Concepts," 154–164; Ryder et al., "Fish Yield Estimator," 663–688; Henderson, Ryder, and Kudhongania, "Assessing Fishery Potentials," 2000–2009.

34. "Minutes of meeting to discuss the proposed establishment of an experimental hatchery to be located in the Sault Ste. Marie district," January 16, 1957, A74–0022, Box 25, University of Toronto Archives.

35. This was the conclusion drawn for the Great Lakes. A different view was taken of fish populations in smaller inland lakes. Fry, for example, concluded that even moderate fishing pressure could readily deplete fish stocks of Opeongo Lake in Algonquin Park and that therefore controls on fishing were necessary (Fry, "Lake Trout Fishery," 27–67). Fry's statistical work on lake trout eventually became the basis for estimating total allowable catches (Whillans, Regier, and Christie, "F. E. J. Fry's Field Studies").

36. "Fishing—Sport or Profit," 1.

37. Huntsman, A. G., to Capt. F. W. Wallace, editor of *Canadian Fisherman*, Febru-

ary 4, 1944, 3, Box 69, Huntsman Collection, B78–0010, University of Toronto Archives.

38. Regier, "Models of Walleye Production"; Cucin and Regier, "Dynamics and Exploitation," 221–274; Regier, Applegate, and Ryder, "Walleye in Western Lake Erie"; Christie, "Effects of Artificial Propagation," 597–639; Christie, "Review of the Changes."

39. Christie commented that the role of biologists in Great Lakes fisheries management has been to "write the obituaries as various prime fish stocks have become history" ("Possible Influences," 32).

40. Goodwin, "Salt Water Trawls," 29–31; Armstrong, "Historical Review," 25–34; Clarke, "Very Uncertain Future," 1.; quote from "Fishing—Sport or Profit," 1.

41. Clarke, "Wildlife in Perspective," 837.

42. Burton, *Natural Resource Policy*, 23–46; Thorpe, "Historical Perspective," 1–13; Dorcey, "Great Expectations," 481–511.

43. Regier, Applegate, and Ryder, "Walleye in Western Lake Erie"; Christie, "Review of the Changes"; Baldwin and Saalfeld, "Commercial Fish Production." The status of the Canadian Great Lakes commercial fisheries can also be followed through accounts in *Canadian Fisherman;* see, e.g., Budden, "Where Are Lake Ontario's Whitefish?"; Budden, "Tragedy of the Great Lakes"; Morris, "Will the Great Lakes Be Fished Out?"; and Morris, "Discouraging Year for the Great Lakes." See also Canada Department of the Environment and Ontario Ministry of Natural Resources, *First report*, 26, 29, 64.

44. Regier and Hartman, "Lake Erie's Fish Community," 1248–1255; McGucken, "Canadian Federal Government"; Ashworth, *Late Great Lakes*.

45. Christie, "Potential of Exotic Fishes," 76; Clarke, "A Very Uncertain Future," 1.

46. Pimlott, "Scientific Activities in Fisheries," 109.

47. Larkin, "Confidential Memorandum," 190; Loftus, "New Approach," 324. These developments paralleled the view emerging among fisheries researchers elsewhere, that management and research focused on achieving maximum yields of individual species often created additional problems, while failing to achieve its objective. See Larkin, "Epitaph," 1–11.

48. "Report of Meeting of Research Programs Committee," March 29, 1971, and "Fisheries Research Priorities," July 8, 1971, both in Ontario Government Archives, Toronto, File RG 1–289–1-12.

49. Research Programs Committee, Meeting, November 7, 1968, RG 1–289–1-12: "Research programs committee," 1968–1972, Ontario Government Archives.

50. Dillon and Rigler, "Simple Method," 1519–1531.

51. Harvey and Coombs, "Physical and Chemical Limnology," 1883–1897.

52. Beamish and Harvey, "La Cloche Mountain Lakes," 1131–1143.

53. Brinkhurst and Chant, *Good, Good Earth.*

54. Regier, "Ecological Considerations."

55. Cameron and Billingsley, *Energy Flow.* The start of the program in 1968, two

years earlier than the American IBP program, reflected the initially greater enthusiasm among Canadian (and European) ecologists, in contrast with the initial reluctance of many American ecologists to become involved (Worthington, *Evolution*).

56. Rigler, "Char Lake Project," 171–198.
57. Regier, Henry A., in conversation with author, July 8, 1992.
58. Bruce, "Water Pollution," 182–193.
59. Regier, "Dynamics of Fish Populations," 3–18; Regier, Henry A., in conversation with author, May 3, 1993.
60. Regier, Applegate, and Ryder, "Walleye in Western Lake Erie."
61. Papers from the symposium were published the following year in Loftus and Regier, "Proceedings of the Symposium," 611–986.
62. Regier, "Sequence of Exploitation," 1992–1999; Regier and Henderson, "Broad Ecological Model," 56–72.
63. Christie, "Lake Ontario," 913–929.
64. Smith, "Species Succession," 667–693.
65. Larkin, "Epitaph," 1–11; Alverson and Paulik, "Objectives and Problems," 1936–1947; Larkin and Wilimovsky, "Contemporary Methods," 1948–1957; Gulland, "New Strategies," 1–11; Carpenter, Kitchell, and Hodgson, "Cascading Trophic Interactions," 634–639; Hilborn, "Living with Uncertainty," 1–5.
66. Canada and Ontario, *First report*, 82.
67. According to this notion, fishermen may reduce a fish stock through excessive fishing. But the increased effort and diminished return implied by a depleted stock will encourage fishermen to exploit other stocks, permitting the depleted stock to restore itself. However, self-regulation is often ineffective because while fishing effort may often be partially diverted to other, usually less valuable fish stocks, enough stress will be maintained on the more valuable species to prevent their recovery.
68. Canada and Ontario, *First report*, 10.
69. Canada and Ontario, *First report*, 4.
70. Canada and Ontario, *First report*, 7–11.
71. Canada Department of the Environment and Ontario Ministry of Natural Resources, *Fourth report*, 6.
72. An accessible summary of the plan is Loftus, Johnson, and Regier, "Federal-Provincial Strategic Planning," 916–927.
73. Canada, Department of the Environment, and Ontario, Ministry of Natural Resources, SPOF Working Group Number Three, *Strategic Planning*, 11.
74. Environment Canada, "Comprehensive River Basin Planning"; Dorcey, "Great Expectations," 486–491.
75. Products of this review include Harvey, "Aquatic Environmental Quality," 2634–2670, and Regier and McCracken, "Shelf-Seas Fisheries," 1887–1932.
76. Fleck, "Restructuring the Ontario Government," 55–68.
77. McEvoy, *Fisherman's Problem*, 227–257.

CHAPTER 8. COMPARING ECOLOGISTS AND THEIR INSTITUTIONS

1. McIntosh, *Background,* 30.
2. Worster, *Nature's Economy,* 293.
3. Mitman, *State of Nature;* Taylor, "Technocratic Optimism"; Kwa, "Representations of Nature"; Kwa, "Modeling the Grasslands."
4. The interpretation in terms of a dichotomy of attitudes or value systems recurs in analyses of environmental politics. O'Riordan, for example, suggests that "the contradictions that beset modern environmentalism reflect the divergent evolution of two ideological themes": the technocentric, which sees the world as a collection of natural resources, and the ecocentric, according to which nature is of intrinsic worth (O'Riordan, *Environmentalism,* 1–19). Consider also Arne Naess's distinction between the "shallow" and the "deep" environmental movements, Jacqueline Cramer and her colleagues' contrast between the "modernist" and "ecological" perspectives, and Anna Bramwell's description of the "environmentalist" and "ecologist" movements (Naess, "Shallow and the Deep," 95–100; Cramer, Eyerman, and Jamison, "Knowledge Interests," 89–115; Bramwell, *20th Century,* 104–105). See also Marx, "Environmental Degradation," 457–465.
5. Petulla, *American Environmentalism,* xii.
6. Worster, *Nature's Economy,* xiii.
7. Dunlap, *Saving America's Wildlife,* x.
8. Croker, *Pioneer Ecologist,* 46–49, quote on 49; Taylor, "Technocratic Optimism."
9. Hagen, *Entangled Bank,* 163; Worster, "Order and Chaos."
10. This also represented an aesthetic response, expressed through a deep affection for the British landscape, and the combined natural and human factors that it represented. When Nicholson traveled to Greenland and the Amazon he found their pristine nature "impersonal," "mindless," and "chilling." The experience reinforced his belief in the value of careful management, to improve on nature (Nicholson, *Environmental Revolution,* 18–26).
11. Rosenberg, "Woods or Trees."
12. Burgess, "United States"; Duff and Lowe, "Great Britain." No similar survey exists for Canadian ecology; neither is there a Canadian society of ecologists to track the growth of a national ecological community. However, the growth of Canadian ecology is implied, for example, in the growth of Canadian membership in the International Association of Limnology, from 63 members in 1964 to almost 200 in 1974 (Harvey and Geen, "Introduction," 503).
13. Dansereau, "Future of Ecology," 20–23.
14. Kingsland, "Elusive Science"; Tobey, *Saving the Prairies.* On universities see Harwood, *Styles,* 157. The difference between success and failure was evident, for example, in the development of ecology in Illinois and Michigan between 1880 and 1910; see Bocking, "Stephen Forbes." More speculatively, the demands of local contexts may have encouraged the formation of distinctive schools of ecology, defined in terms of a particular regional landscape, such as the Prairie

grasslands, or in terms of the receptivity of certain regional institutions to ecology, as in the case of the midwestern universities at which ecology first emerged as a distinct discipline in the United States. An interesting exploration of the significance of local context in shaping disciplinary formation is Pauly, "Appearance of Academic Biology."

15. Woodwell, "Confusion of Paradigms," 8.
16. Levins, "Strategy of Model Building," 421–431.
17. The emergence of the environment as a prominent issue was described as an "environmental revolution" on both sides of the Atlantic; see, e.g., Nicholson, *Environmental Revolution;* Shabecoff, *Fierce Green Fire,* 129–148; and Worster, *Nature's Economy,* 341–342.
18. Nature Conservancy, *Annual Report for year ended 30 September 1964,* 2.
19. As I note elsewhere, this assumption would be eroded by experience.
20. Quote from Green, *Countryside Conservation,* viii; Heidenheimer, Heclo, and Adams, "Environmental Policy," 324; Vogel, *National Styles.*
21. Balogh, *Chain Reaction.*
22. Harwood, *Styles,* 178; see also Kohler, *Medical Chemistry.*
23. On the Nature Conservancy see Max Nicholson, "Open-Air Laboratories," *The Times,* August 21, 1954. On Oak Ridge see ORNL, *Health Physics Report, 1961,* 105–106. "Experimental limnology" is how Likens's teacher Arthur Hasler described his own approach. It served as one inspiration for Bormann and Likens's experimental approach to ecosystems (Hasler, "Experimental Limnology," 36). On Ontario see Fry, "South Bay Experiment."
24. In fact this invocation of scientific autonomy obscured the focus of postwar science, particularly in the United States, on the priorities of the national security state. In Canada and Great Britain this focus was less evident; efficient natural resource and land-use management were of chief concern.
25. On the debate among British ecologists see Sheail, *Seventy-Five Years,* 224–239; among ecologists generally see McIntosh, *Background,* 289–323.
26. Anderson, "Policy Determination."
27. Shabecoff, *Fierce Green Fire,* 91; Udall, "Message for Biologists," 17–18; Caldwell, *Man and His Environment,* 30–31.
28. U.S. House, *International Biological Program,* 2.
29. Chisholm, *Philosophers,* xi.
30. Kingsland, "Elusive Science," 177.
31. U.S. House, *Environmental Pollution,* 5–6.
32. Steinhart and Cherniack, *Universities and Environmental Quality,* 3.
33. Caldwell, *Challenge,* 175. Note also President's Science Advisory Committee, *Restoring,* 196–197.
34. These critiques also provoked substantial reaction, as Rachel Carson found. *Silent Spring* had its impact through its specific evidence and its underlying argument that pesticide regulation was too important to be left to the iron triangle of farmers, the chemical industry, and the U.S. Department of Agriculture and its

experts. Vociferous criticisms of her from these parties addressed this directly, arguing that she did not have the credentials to contribute to the issue. Indeed, in corporatist terms, she didn't. The point, however, was that these terms were no longer the only ones that were relevant.

35. On Environment Canada see Doern and Conway, *Greening of Canada*, 16–20.

36. U.S. House, *Environmental Pollution*, 4; Averch, *Strategic Analysis*, 121–122; Caldwell, "New Focus," 132–139.

37. Brickman, "Toxic Chemical Regulation"; Webster, *Science;* Vogel, *National Styles;* Hays, *Beauty, Health, and Permanence*, 330–333; Nelkin, "Political Conflict"; Heidenheimer, Heclo, and Adams, "Environmental Policy," 323. On confrontation in British environmental politics see Lowe et al., *Countryside Conflicts;* Shoard, *Theft of the Countryside*. One reflection of this preference for confrontation is recent criticism of the conservancy's reliance on science, communication, and consensus as ineffectual against agricultural activities that conflict fundamentally with conservation objectives; see, e.g., Pye-Smith and Rose, *Crisis and Conservation*.

38. Adam, "Environmental Assessment"; Paelhke and Torgerson, *Managing Leviathan*.

39. On Great Britain see Ronayne, *Science in Government;* Vig, "Policies." For nature conservation and ecology, the chief implication of this approach was apparent in the arrangement between the Nature Conservancy Council and the Institute for Terrestrial Ecology: part of the Institute's funds would be obtained through contracts placed by the Council and other agencies (Sheail, *Nature in Trust*, 196–243). On Canada see Dufour and de la Mothe, "Historical Conditioning"; Phillipson, "National Research Council." On the United States see Lakoff, "Congress"; Morison, "Problems"; Averch, *Strategic Analysis*, 20–25; Stine, *History*.

40. Jasanoff, "Renegotiation"; Webster, *Science*, 51–52; Jeffrey, "Adversarial Arena."

41. Because my examination of Nature Conservancy research extends only to the early 1960s, this case will not be discussed specifically.

42. Suter, "NEPA Assessment."

43. White, "Environment," 186.

44. Nelkin, "Professional Responsibility," 84. A major 1976 symposium on environmental modeling, for example, was dominated by scientists not identifying themselves as ecologists (Ott, *Environmental Modeling and Simulation*). On Bodega Bay see Balogh, *Chain Reaction*, 240.

45. Slobodkin, "Intellectual Problems," 338.

46. See, e.g., Lowrance, *Acceptable Risk*.

47. Schumacher, *Small Is Beautiful;* Lovins, *Soft Energy Path;* Hays, *Beauty, Health, and Permanence*, 250–251.

48. Hays, *Beauty, Health, and Permanence*, 76–79; Gelpe and Tarlock, "Uses of Scientific Information," 371–427.

49. Shrader-Frechette and McCoy, in *Method in Ecology*, provide an extensive dis-

cussion of these aspects of ecology and their implications for ecology's role in environmental problem solving.

50. See, e.g., Ehrlich, "Ecology and Resource Management."

51. White, "Role of Scientific Information"; Loucks, "Role of Basic Ecological Knowledge"; Brooks, "Science Indicators."

52. White, "Environment," 189.

53. By the time they wrote their second book on the Hubbard Brook ecosystem, published in 1979, Bormann and Likens could report that several of their results had been incorporated into forestry practices.

54. Weiss, "Policy's Sake"; Sabatier and Jenkins-Smith, *Policy Change*.

55. This has an implication for historical method: if one wishes to identify the current priorities of a discipline, the work of not only its more senior members should be examined, but also that of their students, because their immediate prospects depend on reading these priorities correctly.

56. Adam, "Environmental Assessment"; Weston, *Federal Environmental Assessment*, 5.

CHAPTER 9. UNDERSTANDING ECOLOGY'S PLACE IN THE WORLD

1. On contingency in environmental history see esp. Cronon, "Environmental History," 10–16.

2. Ozawa, *Recasting Science;* Leiss and Chociolko, *Risk and Responsibility,* esp. 197–218. For a case study of the role of epistemic communities in developing consensus on comprehensive approaches to environmental protection see Haas, *Saving the Mediterranean,* esp. 214–247. For an example of an effort to encourage consensus on protection of a regional ecosystem see Royal Commission on the Future of the Toronto Waterfront, *Regeneration.*

3. Cronon, "Environmental History," 20.

References

MANUSCRIPT SOURCES

Stanley Auerbach, personal files, Oak Ridge National Laboratory, Oak Ridge, Tennessee.

Eastern Deciduous Forest Biome File, Environmental Sciences Division Library, Oak Ridge National Laboratory, Oak Ridge, Tennessee.

Hubbard Brook Ecosystem Study Files, Institute of Ecosystem Studies, New York Botanical Garden, Mary Flagler Cary Arboretum, Millbrook, New York.

Nature Reserves Investigation Committee, File HLG 92–18; National Parks Bill, 1948–1949, Functions and Responsibilities of Nature Conservancy Board, File HLG 92–180; National Parks Committee, "Wild Birds: Establishment of Special Nature Reserve: Correspondence with the Home Office, etc.," 1945–1948, File HLG 93–29; National Parks Committee, "Correspondence with the Secretary of the NRIC," File HLG 93–39; Wildlife Conservation Special Committee, "Terms of Reference, Committee Papers, Minutes of Meetings, etc.," File HLG 93–48; WLCSC, "Reports 1–18," File HLG 93–51; WLCSC, "Reports 20–25," File HLG 93–52; "The Nature Conservancy: Agricultural Liaison in England and Wales," File

MAF 141–191; Nature Conservancy: Minutes of Meetings, File MAF 131–25; Advisory Council on Scientific Policy, File CAB 132–66. In the Public Record Office, London.

Henry A. Regier, personal files, University of Toronto.

Department of Zoology papers, File A74–0022; Frederick E. J. Fry papers, File B88–0027; A. G. Huntsman papers, File B78–0010. University of Toronto Archives, Toronto.

Advisory Committee for Harkness Laboratory, 1966, File RG 1–289-1-2; Advisory Committee for Harkness Laboratory, 1967–1972, File RG 1–289-1-3; Fisheries Research Advisory Committee, 1966–1972, File RB 1–289-1-8; Research Programs Committee, 1968–1972, File RG 1–289-1-12; Lake Survey Procedures, File RG 1–282-1-24. Ontario Government Archives, Toronto.

Adam, Marie P. "Environmental Assessment and Review Processes in Canada." In *The Role of Environmental Impact Assessment in the Decisionmaking Process; Proceedings of an International Workshop, August 1987*, 119–129. Berlin: Erich Schmidt Verlag, 1989.

Allee, W. C., Alfred E. Emerson, Orlando Park, Thomas Park, and Karl P. Schmidt. *Principles of Animal Ecology*. Philadelphia: W. B. Saunders, 1949.

Allen, David E. *The Naturalist in Britain: A Social History*. London: Allen Lane, 1976.

———. "Changing Attitudes to Nature Conservation: The Botanical Perspective." *Biological Journal of the Linnean Society* 32 (1987): 203–212.

Alverson, D. L., and G. J. Paulik. "Objectives and Problems of Managing Aquatic Living Resources." *Journal of the Fisheries Research Board of Canada* 30 (1973): 1936–1947.

Anderson, Frances. "Policy Determination of Government Scientific Organizations: A Case Study of the Fisheries Research Board of Canada, 1963–1973." Ph.D. diss., University of Montreal, 1988.

Anderson, Roger. "Environmental, Safety, and Health Issues at U.S. Nuclear Weapons Production Facilities, 1946–1988." *Environmental Review* 13 (1989): 69–92.

Armstrong, G. C. "An Historical Review of the Management of the Sport Fishery in Ontario." *Ontario Fish and Wildlife Review* 6, nos. 1–2 (1967): 25–34.

Ashworth, William. *The Late Great Lakes: An Environmental History*. Toronto: Collins, 1986.

Auerbach, Stanley I. "The Soil Ecosystem and Radioactive Waste to the Ground." *Ecology* 39 (1958): 522–529.

———. "Long-Term Role of ORNL Ecology Program." February 1962. Typescript.

———. "IBP Biome Research Proposal to the Environmental Biology Program, National Science Foundation." Oak Ridge National Laboratory, 1969.

———. "Analysis of the Structure and Function of Ecosystems in the Deciduous Forest Biome." Proposal to the National Science Foundation for 1 July 1970–30 June 1971. Oak Ridge National Laboratory, 1970.

———. "The Deciduous Forest Biome Program in the United States of America." In

Productivity of Forest Ecosystems. Proceedings of the Brussels Symposium, 27–31 October 1969, ed. P. Duvigneaud, 677–684. Paris: Unesco, 1971.

———. "Ecology, Ecologists, and the E. S. A." Oak Ridge National Laboratory, 1972. Photocopy.

———. "Ecology's Twenty-Year Space Program." *Oak Ridge National Laboratory Review* 8 (1975): 21–25.

———. "Current Perceptions and Applicability of Ecosystem Analysis to Impact Assessment." *Ohio Journal of Science* 78 (1978): 163–174.

———. "Ecosystem Response to Stress: A Review of Concepts and Approaches." In *Stress Effects on Natural Ecosystems*, ed. Gary W. Barrett and Rutger Rosenberg, 29–41. Chichester: John Wiley and Sons, 1981.

———. "Evolution of ORNL's Environmental Sciences." *Oak Ridge National Laboratory Review* 22, nos. 2–3 (1989): 1–9.

Auerbach, Stanley I., and D. A. Crossley. "Strontium-90 and Cesium-137 Uptake by Vegetation under Natural Conditions." *Proceedings of the Second International Conference on Peaceful Uses of Atomic Energy* 18 (1958): 494–499.

Auerbach, Stanley I., Steven V. Kaye, Daniel J. Nelson, David E. Reichle, Paul B. Dunaway, and Ray S. Booth. "Understanding the Dynamic Behavior of Radionuclides Released to the Environment and Implications." In *Proceedings of the Fourth United Nations International Conference on the Peaceful Uses of Atomic Energy, Sept. 6–16, 1971*, 3.3–1-3.3.18.

Auerbach, Stanley I., Daniel J. Nelson, Steven V. Kaye, David E. Reichle, and Charles C. Coutant. "Ecological Considerations in Reactor Power Plant Siting." *Proceedings, International Atomic Energy Agency Symposium 146/53* (1971): 804–820.

Auerbach, Stanley I., and Jerry S. Olson. "Biological and Environmental Behavior of Ruthenium and Rhodium." In *Radioecology: Proceedings of the First National Symposium on Radioecology, September 10–15, 1961*, ed. Vincent Schultz and Alfred W. Klement, 509–519. New York: Reinhold, 1963.

Auerbach, Stanley I., Jerry S. Olson, and H. D. Waller. "Landscape Investigations Using Caesium-137." *Nature* 201 (1964): 761–764.

Auerbach, Stanley I., and Vincent Schultz. *Onsite Ecological Research of the Division of Biology and Medicine at the Oak Ridge National Laboratory*. U.S. Atomic Energy Commission, TID-16890, 1962.

Averch, Harvey A. *A Strategic Analysis of Science and Technology Policy*. Baltimore: Johns Hopkins University Press, 1985.

Baker, John R. *Sir Arthur Tansley, 1871–1955*. Occasional Pamplet no. 16. London: Society for Freedom in Science, 1955.

Baldwin, Henry I., and Charles F. Brooks. *Forests and Floods in New Hampshire*. Boston: New England Regional Planning Commission, 1936.

Baldwin, Norman S., and Robert W. Saalfeld. "Commercial Fish Production in the Great Lakes, 1867–1960." *Great Lakes Fishery Commission Technical Report* no. 3, 1962.

Balogh, Brian. *Chain Reaction: Expert Debate and Public Participation in American*

Commercial Nuclear Power, 1945–1975. Cambridge: Cambridge University Press, 1991.

Barnthouse, Lawrence W., John Boreman, Sigurd W. Christensen, C. P. Goodyear, Webster Van Winkle, and Douglas S. Vaughan. "Population Biology in the Courtroom: The Hudson River Controversy." *BioScience* 34, no. 1 (1984): 14–19.

Barnthouse, Lawrence W., John Boreman, Thomas L. Englert, William L. Kirk, and Edward G. Horn. "Hudson River Settlement Agreement: Technical Rationale and Cost Considerations." *American Fisheries Society Monograph* 4 (1988): 267–273.

Barnthouse, Lawrence W., Ronald J. Klauda, and Douglas S. Vaughan. "Introduction to the Monograph." *American Fisheries Society Monograph* 4 (1988): 1–8.

Battelle Columbus Laboratories. *Evaluation of Three of the Biome Studies Programs Funded Under the Foundation's International Biological Program (IBP). Final Report*. Columbus, Ohio: Battelle, 1975.

Beamish, Richard J., and Harold H. Harvey. "Acidification of the La Cloche Mountain Lakes, Ontario, and Resulting Fish Mortalities." *Journal of the Fisheries Research Board of Canada* 29 (1972): 1131–1143.

Beckel, Annamarie L. "Breaking New Waters: A Century of Limnology at the University of Wisconsin." *Transactions of the Wisconsin Academy of Sciences, Arts and Letters*, Special Issue, 1987.

Bensley, B. A. "A Plan for the Biological Investigation of the Water Areas of Ontario." *University of Toronto Studies* 1 (1922): 6–7, 13–20.

Benson, Norman G. "The American Fisheries Society, 1920–1970." In *A Century of Fisheries in North America*, ed. Norman G. Benson, 13–24. Washington, D.C.: American Fisheries Society, 1970.

Bernal, John D. *The Social Function of Science*. London: Routledge, 1939.

Bocking, Stephen. "Stephen Forbes, Jacob Reighard and the Emergence of Aquatic Ecology in the Great Lakes Region." *Journal of the History of Biology* 23 (1990): 461–498.

Boffey, Philip M. "Radioactive Pollution: Minnesota Finds AEC Standards Too Lax." *Science* 163 (1969): 1043–1046.

Booth, Ray S., and Stephen V. Kaye. "A Preliminary Systems Analysis Model of Radioactivity Transfer to Man from Deposition in a Terrestrial Environment." Oak Ridge National Laboratory, ORNL/TM-3135, 1971.

Booth, Ray S., Stephen V. Kaye, and Paul S. Rohwer. "A Systems Analysis Methodology for Predicting Dose to Man from a Radioactively Contaminated Terrestrial Environment." In *Radionuclides in Ecosystems: Proceedings of the Third National Symposium on Radioecology*, ed. Daniel J. Nelson, 877–893. U.S. Atomic Energy Commission, CONF-710501, 1972.

Bormann, F. Herbert. "Factors Determining the Role of Loblolly Pine and Sweetgum in Early Old-Field Succession in the Piedmont of North Carolina." *Ecological Monographs* 23 (1953): 339–358.

———. "The Primary Leaf as an Indicator of Physiologic Condition in Shortleaf Pine." *Forest Science* 1 (1955): 189–192.

————. "Percentage Light Readings, Their Intensity-Duration Aspects, and Their Significance in Estimating Photosynthesis." *Ecology* 37 (1956): 473–476.

————. "Moisture Transfer Between Plants Through Intertwined Root Systems." *Plant Physiology* 32 (1957): 48–55.

————. "An Inseparable Linkage: Conservation of Natural Ecosystems and the Conservation of Fossil Energy." *BioScience* 26 (1976): 754–760.

————. "The New England Landscape: Air Pollution Stress and Energy Policy." *Ambio* 11 (1982): 188–194.

————. "The Effects of Air Pollution on the New England Landscape." *Ambio* 11 (1982): 338–346.

————. "Lessons From Hubbard Brook." In *Proceedings of the Chaparral Ecosystems Research Meeting,* ed. E. Keller, S. Cooper, and J. De Vries. Report 62. California Water Resources Center, University of California, Davis, 1986.

Bormann, F. Herbert, and Ben F. Graham. "The Occurrence of Natural Root Grafting in Easter White Pine, *Pinus strobus* L., and Its Ecological Implications." *Ecology* 40 (1959): 677–691.

Bormann, F. Herbert, and Gene E. Likens. "Nutrient Cycling." *Science* 155 (1967): 424–429.

————. "The Watershed-Ecosystem Concept and Studies of Nutrient Cycles." In *The Ecosystem Concept in Natural Resource Management,* ed. George M. Van Dyne, 49–76. New York: Academic Press, 1969.

————. "Hydrologic-Mineral Cycle Interaction in Small Undisturbed and Man-Manipulated Ecosystems (Watersheds)." Proposal to the Environmental Biology Program, National Science Foundation, to start 1 September 1969, for two years.

————. "Hydrologic-Nutrient Cycle Interaction in Small Undisturbed and Man-Manipulated Ecosystems." Proposal to the Ecosystem Studies Program, National Science Foundation, to start 1 September 1976, for two years.

————. "The Fresh Air-Clean Water Exchange." *Natural History* 86, no. 9 (1977): 62–71.

————. "Hydrologic-Nutrient Cycle Interaction in Small Disturbed and Man-Manipulated Ecosystems." Proposal to the National Science Foundation, to start 1 September 1978, for three years.

————. *Pattern and Process in a Forested Ecosystem: Disturbance, Development and the Steady State Based on the Hubbard Brook Ecosystem Study.* New York: Springer-Verlag, 1979.

Bormann, F. Herbert, Gene E. Likens, and John S. Eaton. "Biotic Regulation of Particulate and Solution Losses from a Forest Ecosystem." *BioScience* 19 (1969): 600–610.

Bormann, F. Herbert, Gene E. Likens, D. W. Fisher, and Robert S. Pierce. "Nutrient Loss Accelerated By Clear-Cutting of a Forest Ecosystem." *Science* 159 (1968): 882–884.

Bormann, F. Herbert, Gene E. Likens, and Noye M. Johnson. "Hydrologic-Mineral Cycle Interaction in a Small Watershed." Proposal to the U.S. Public Health Service, National Institutes of Health, 1963.

Bormann, F. Herbert, Gene E. Likens, Thomas G. Siccama, Robert S. Pierce, and John S. Eaton. "The Export of Nutrients and Recovery of Stable Conditions Following Deforestation at Hubbard Brook." *Ecological Monographs* 44 (1974): 255–277.

Bormann, F. Herbert, Thomas G. Siccama, Gene E. Likens, and Robert H. Whittaker. "The Hubbard Brook Ecosystem Study: Composition and Dynamics of the Tree Stratum." *Ecological Monographs* 40 (1970): 373–388.

Botkin, Daniel B. "Forests, Lakes, and the Anthropogenic Production of Carbon Dioxide." *BioScience* 27 (1977): 325–331.

Botkin, Daniel B., James F. Janak, and James R. Wallis. "Some Ecological Consequences of a Computer Model of Forest Growth." *Journal of Ecology* 60 (1972): 849–872.

Bramwell, Anna. *Ecology in the Twentieth Century: A History*. New Haven: Yale University Press, 1989.

Brickman, Ronald. "Science and the Politics of Toxic Chemical Regulation: U.S. and European Contrasts." *Science, Technology, and Human Values* 9 (1984): 107–111.

Brinkhurst, Ralph O., and Donald A. Chant. *This Good, Good Earth: Our Fight for Survival*. Toronto: Macmillan of Canada, 1971.

British Ecological Society. *Nature Conservation and Nature Reserves*. Cambridge: Cambridge University Press, 1943.

———. "Ecological Principles Involved in the Practice of Forestry." *Journal of Ecology* 32 (1944): 83–115.

———. "Joint Meeting with the British Society of Soil Science, Held in the Botany Department, University College, London, on 28 April 1948." *Journal of Ecology* 36 (1948): 328.

———. "Spring and Summer Meetings, 1956." *Journal of Ecology* 45 (1957): 604–611.

Bronstein, Daniel. "The AEC Decision-Making Process and the Environment: A Case Study of the Calvert Cliffs Nuclear Power Plant." *Ecology Law Quarterly* 1 (1971): 689–721.

Brooks, Harvey. *The Government of Science*. Cambridge: MIT Press, 1968.

Brooks, Harvey. "Science Indicators and Science Priorities." *Science, Technology and Human Values* 7 (1982): 14–31.

Bruce, J. P. "Water Pollution and the Role of the Canada Centre for Inland Waters." *Canadian Geographical Journal* 80, no. 6 (1970): 182–193.

Brungs, W. A. "Effects of Heated Water from Nuclear Plants on Aquatic Life." In *Nuclear Power and the Public,* ed. H. Foreman. Minneapolis: University of Minnesota Press, 1970.

Budd, John. "Movements of Tagged Whitefish in Northern Lake Huron and Georgian Bay." *Transactions of the American Fisheries Society* 86 (1957): 128–134.

Budden, F. R. "Where Are Lake Ontario's Whitefish?" *Canadian Fisherman,* May 1953, 25.

———. "The Tragedy of the Great Lakes." *Canadian Fisherman,* April 1954, 29–31.

Bukharin, N. I., et al. *Science at the Crossroads: Papers Presented to the International*

Congress of the History of Science and Technology, London, 29 June–3 July 1931, by the Delegates of the U.S.S.R. London: Kniga, 1931.

Bunnell, Fred. "Theological Ecology or Models and the Real World." *Forestry Chronicle* 49 (1973): 167–171.

Burgess, Robert L. "United States." In *Handbook of Contemporary Developments in World Ecology*, ed. Edward J. Kormondy and Frank J. McCormick, 67–101. Westport, Conn.: Greenwood, 1981.

Burgess, Robert L., and Stanley I. Auerbach. "Objectives, Strategies, and Operational Management of an Integrated Research Program—The Eastern Deciduous Forest Biome." EDFB Memo Report 71–21, 1971.

Burgess, Robert L., and L. H. Kern. "Progress Report, 1971–72: Eastern Deciduous Forest Biome, US-IBP." ORNL-IBP Report 73–5, 1973.

Burgess, Robert L., and Robert V. O'Neill. "Eastern Deciduous Forest Biome Progress Report and Continuation Proposal." ORNL-IBP Report 74–1, 1974.

Burton, Thomas L. *Natural Resource Policy in Canada: Issues and Perspectives.* Toronto: McClelland and Stewart, 1972.

Cairns, John. "We're in Hot Water." *Scientist and Citizen,* October 1968, 10.

Caldwell, Lynton K. "Environment: A New Focus for Public Policy?" *Public Administration Review* 23 (1963): 132–139.

———. *Environment: A Challenge for Modern Society.* New York: Natural History Press, 1970.

———. *Man and His Environment: Policy and Administration.* New York: Harper and Row, 1975.

Cameron, Thomas W. M., and L. W. Billingsley, eds. *Energy Flow—Its Biological Dimensions: A Summary of the IBP in Canada, 1964–1974.* Ottawa: Royal Society of Canada, 1975.

Canada, Department of the Environment. *Monograph on Comprehensive River Basin Planning.* Ottawa: Information Canada, 1975.

Canada, Department of the Environment, and Ontario, Ministry of Natural Resources. *First Report of the Federal-Provincial Working Group on Strategic Planning for Ontario Fisheries: Preliminary Analysis of Goals and Issues.* October 1974.

Canada, Department of the Environment, and Ontario, Ministry of Natural Resources. *Fourth Report, Federal-Provincial Strategic Planning for Ontario Fisheries: Management Strategies for the 1980s.* April 1976.

Canada, Department of the Environment, and Ontario, Ministry of Natural Resources, SPOF Working Group Number Three. *Strategic Planning for Ontario Fisheries, Policy Development: Experimental Management.* October 1978.

Capshew, James H., and Karen A. Rader. "Big Science: Price to the Present." *Osiris,* 2nd ser., 7 (1992): 3–25.

Carlisle, A., A. H. F. Brown, and E. J. White. "Litter Fall, Leaf Production and the Effects of Defoliation by Tortrix Viridana in a Sessile Oak (*Quercus petraea*) Woodland." *Journal of Ecology* 54 (1966): 65–76.

———. "The Nutrient Content of Tree Stem Flow and Ground Flora Litter and

Leachates in a Sessile Oak (*Quercus petraea*) Woodland." *Journal of Ecology* 55 (1967): 615–627.

Carpenter, Stephen R., James F. Kitchell, and James R. Hodgson. "Cascading Trophic Interactions and Lake Productivity." *BioScience* 35 (1985): 634–639.

Champion, H. G. "The Future of Our New Forests." *Advancement of Science* 6 (1950): 319–324.

Cherry, Gordon E. *Environmental Planning, 1939–1969*. London: HMSO, 1975.

Chisholm, Anne. *Philosophers of the Earth—Conversations with Ecologists*. London: Sidgwick and Jackson, 1972.

Christensen, Sigurd W., and Thomas L. Englert. "Historical Development of Entrainment Models for Hudson River Striped Bass." *American Fisheries Society Monograph* 4 (1988): 133–142.

Christensen, Sigurd W., Webster Van Winkle, and Jack S. Mattice. "Defining and Determining the Significance of Impacts: Concepts and Methods." In *Proceedings of the Workshop on the Biological Significance of Environmental Impacts*, ed. R. K. Sharma et al., 191–219. Washington, D.C.: Nuclear Regulatory Commission, NR-CONF-002, 1976.

Christie, W. Jack "Effects of Artificial Propagation and the Weather on Recruitment in the Lake Ontario Whitefish Fishery." *Journal of the Fisheries Research Board of Canada* 20 (1963): 597–639.

———. "The Potential of Exotic Fishes in the Great Lakes." In "A Symposium on Introductions of Exotic Species," ed. Kenneth H. Loftus, 73–91. *Ontario Department of Lands and Forests Report*, no. 82, 1968.

———. "Possible Influences of Fishing in the Decline of Great Lakes Fish Stocks." *Proceedings of the 11th Conference on Great Lakes Research* (1968): 31–38.

———. "Lake Ontario: Effects of Exploitation, Introductions, and Eutrophication on the Salmonid Community." *Journal of the Fisheries Research Board of Canada* 29 (1972): 913–929.

———. "A Review of the Changes in the Fish Species Composition of Lake Ontario." *Great Lakes Fishery Commission Technical Report*, no. 23, 1973.

Clapham, A. R. "William Harold Pearsall, 1891–1964." *Biographical Memoirs of the Fellows of the Royal Society* 17 (1971): 511–540.

Clark, John R. "Thermal Pollution and Aquatic Life." *Scientific American* 220, no. 3 (1969): 18–27.

Clarke, C. H. D. "Wildlife in Perspective." In *Resources for Tomorrow*, 2: 837–844. Ottawa: Queen's Printer, 1961.

———. "A Very Uncertain Future." *Ontario Fish and Wildlife Review* 9, nos. 1–2 (1970): 1.

Clemens, Wilbert A. *Education and Fish*. Fisheries Research Board of Canada, Manuscript Report Series, no. 974, May 1968.

Coble, Daniel W. "The White Sucker Population of South Bay, Lake Huron, and Effects of the Sea Lamprey on It." *Journal of the Fisheries Research Board of Canada* 24 (1967): 2117–2136.

Colman, E. A. *Vegetation and Watershed Management*. New York: Ronald Press, 1953.

Connery, Robert H. *Governmental Problems in Wild Life Conservation*. New York: Columbia University Press, 1935.

Conway, Verona M. "Lines of Development in Ecology." *Proceedings of the Chemical Society* (1957): 246–250.

Cook, Robert E. "Raymond Lindeman and the Trophic-Dynamic Concept in Ecology." *Science* 198 (1977): 22–26.

Cowser, K. E., and F. L. Parker. "Soil Disposal of Radioactive Wastes at ORNL: Criteria and Techniques of Site Selection and Monitoring." *Health Physics* 1 (1958): 152–163.

Cramer, Jacqueline, Ron Eyerman, and Andrew Jamison. "The Knowledge Interests of the Environmental Movement and Its Potential for Influencing the Development of Science." In *The Social Direction of the Public Sciences, Sociology of the Sciences Yearbook*, vol. 11, ed. S. Blume et al., 89–115. Dordrecht, Netherlands: D. Reidel, 1987.

Croker, Robert A. *Pioneer Ecologist: The Life and Work of Victor Ernest Shelford, 1877–1968*. Washington, D.C.: Smithsonian Institution Press, 1991.

Cronon, William. "The Uses of Environmental History." *Environmental History Review* 17, no. 3 (1993): 1–22.

Crossley, D. A. "Movement and Accumulation of Radiostrontium and Radiocesium in Insects." In *Radioecology: Proceedings of the First National Symposium on Radioecology, September 10–15, 1961*, ed. Vincent Schultz and Alfred W. Klement, 103–105. New York: Reinhold, 1963.

———. "Consumption of Vegetation by Insects." In *Radioecology: Proceedings of the First National Symposium on Radioecology, September 10–15, 1961*, ed. Vincent Schultz and Alfred W. Klement, 427–430. New York: Reinhold, 1963.

———. "Use of Radioactive Tracers in the Study of Insect-Plant Relationships." In *Radiation and Radioisotopes Applied to Insects of Agricultural Importance*, 43–53. Vienna: International Atomic Energy Agency, 1963.

Crossley, D. A., and Henry F. Howden. "Insect-Vegetation Relationships in an Area Contaminated by Radioactive Wastes." *Ecology* 42 (1961): 302–317.

Crossley, D. A., and David E. Reichle. "Analysis of Transient Behavior of Radioisotopes in Insect Food Chains." *BioScience* 19 (1969): 341–343.

Crowcroft, Peter. *Elton's Ecologists: A History of the Bureau of Animal Population*. Chicago: University of Chicago Press, 1991.

Cucin, Daniel, and Henry A. Regier. "Dynamics and Exploitation of Lake Whitefish in Southern Georgian Bay." *Journal of the Fisheries Research Board of Canada* 23 (1965): 221–274.

Curlin, James W., and Daniel J. Nelson. "Walker Branch Watershed Project: Objectives, Facilities, and Ecological Characteristics." Oak Ridge National Laboratory, ORNL/TM-2271, 1968.

Dansereau, Pierre. "The Future of Ecology." *BioScience* 14, no. 7 (1964): 20–23.

Dawson, Frank G. *Nuclear Power: Development and Management of a Technology*. Seattle: University of Washington Press, 1976.

Del Sesto, Steven L. *Science, Politics, and Controversy: Civilian Nuclear Power in the United States, 1946–1974*. Boulder, Colo.: Westview, 1979.

Dickson, David. *The New Politics of Science*. Chicago: University of Chicago Press, 1984.

Dillon, Peter J., and Frank H. Rigler. "A Simple Method for Predicting the Capacity of a Lake for Development Based on Lake Trophic Status." *Journal of the Fisheries Research Board of Canada* 32 (1975): 1519–1531.

Doern, G. Bruce. *Science and Politics in Canada*. Montreal: McGill–Queen's University Press, 1972.

Doern, G. Bruce, and Thomas Conway. *The Greening of Canada: Federal Institutions and Decisions*. Toronto: University of Toronto Press, 1994.

Dominski, A. S. "Accelerated Nitrate Production and Loss in the Northern Hardwood Forest Ecosystem Underlain by Podzol Soils Following Clear Cutting and Addition of Herbicides." Ph.D. diss., Yale University, 1971.

Dorcey, Anthony H. J. "Research for Water Resources Management: The Rise and Fall of Great Expectations." In *Canadian Aquatic Resources*, ed. Michael C. Healey and Ron R. Wallace, 481–511. *Canadian Bulletin of Fisheries and Aquatic Sciences*, 215 (1987).

Dower, John. *National Parks in England and Wales*. London, 1945.

Duff, Andrew G., and Philip D. Lowe. "Great Britain." In *Handbook of Contemporary Developments in World Ecology*, ed. Edward J. Kormondy and Frank J. McCormick, 141–156. Westport, Conn.: Greenwood, 1981.

Dufour, Paul, and John de la Mothe. "The Historical Conditioning of S&T." In *Science and Technology in Canada*, ed. Paul Dufour and John de la Mothe, 6–22. Harlow, Essex: Longman, 1993.

Dunlap, Thomas R. *Saving America's Wildlife: How Science Changed Our Minds*. Princeton: Princeton University Press, 1988.

Duvigneaud, P., ed. *Productivity of Forest Ecosystems. Proceedings of the Brussels Symposium, 27–31 October 1969*. Paris: Unesco, 1971.

Dymond, John R. "The Problems of Fisheries and Wildlife Research in Ontario." Ontario Fisheries Research Laboratory, 1951. Mimeograph.

———. "Artificial Propagation in the Management of Great Lakes Fisheries." *Transactions of the American Fisheries Society* 86 (1957): 384–392.

———. "Advisory Committee on Fisheries and Wildlife Research." Ontario Research Foundation, 1959. Mimeograph. Harkness Library, University of Toronto, Department of Zoology.

———. "The Fisheries of the Great Lakes." In *Fish and Wildlife: A Memorial to W. J. K. Harkness*, ed. John R. Dymond, 87. Toronto: Longmans, 1964.

Dymond, John R., and A. V. Delaporte. "Pollution of the Spanish River." *Ontario Department of Lands and Forests Research Paper* no. 25, 1952.

Eberhardt, L. L. "Quantitative Ecology and Impact Assessment." *Journal of Environmental Management* 4 (1976): 27–70.

Ehrlich, Paul R. "Discussion: Ecology and Resource Management—Is Ecological Theory any Good in Practice?" In *Perspectives in Ecological Theory*, ed. Jonathan Roughgarden, Robert M. May, and Simon A. Levin, 306–318. Princeton: Princeton University Press, 1989.

Elton, Charles. *Animal Ecology*. London: Sidgwick and Jackson, 1927.

———. "Land Ecology." *Journal of Animal Ecology* 14 (1945): 53.

———. "Population Interspersion: An Essay on Animal Community Patterns." *Journal of Ecology* 37 (1949): 1–23.

Elton, Charles, and Richard S. Miller. "The Ecological Survey of Animal Communities: With a Practical System of Classifying Habitats by Structural Characters." *Journal of Ecology* 42 (1954): 460–496.

Englert, Thomas L., and John Boreman. "Historical Review of Entrainment Impact Estimates and the Factors Influencing Them." *American Fisheries Society Monograph* 4 (1988): 143–151.

Executive Office of the President, Office of Science and Technology. *The Universities and Environmental Quality—Commitment to Problem Focused Education*, by John S. Steinhart, and Stacie Cherniack. Washington, D.C.: Government Printing Office, 1969.

Federer, C. Anthony. "A Soil-Plant-Atmosphere Model for Transpiration and Availability of Soil Water." *Water Resources Research* 15 (1979): 555–562.

Fisher, Stuart G., and Gene E. Likens. "Stream Ecosystem: Organic Energy Budget." *BioScience* 22 (1972): 33–35.

———. "Energy Flow in Bear Brook, New Hampshire: An Integrative Approach to Stream Ecosystem Metabolism." *Ecological Monographs* 43 (1973): 421–439.

"Fishing—Sport or Profit." *Ontario Fish and Wildlife Review* 5, no. 2 (1966): 1.

Fleck, James D. "Restructuring the Ontario Government." *Canadian Public Administration* 16 (1973): 55–68.

Foin, Theodore C. "Systems Ecology and the Future of Human Society." In *Systems Analysis and Simulation in Ecology*, ed. Bernard C. Patten, 2: 475–531. New York: Academic Press, 1972.

Forcier, L. K., and F. Herbert Bormann. "An Intraspecific Energy Flow Model for Tree Populations in Mature Forests." *Bulletin of the Ecological Society of America* 50, no. 2 (1969): 92–93.

Forman, Paul. "Behind Quantum Electronics: National Security as Basis for Physical Research in the United States, 1940–1960." *Historical Studies in the Physical and Biological Sciences* 18 (1987): 149–229.

Fraser, J. M. "The Smallmouth Bass Fishery of South Bay, Lake Huron." *Journal of the Fisheries Research Board of Canada* 12 (1955): 147–177.

Fry, Fred E. "The South Bay Experiment." *Sylva* 3, no. 6 (1947): 15–25.

———. "Effects of the Environment on Animal Activity." *University of Toronto Studies* 55 (1947): 5–62.

———. "Statistics of a Lake Trout Fishery." *Biometrics* 5 (1949): 27–67.

———. "The 1944 Year Class of Lake Trout in South Bay, Lake Huron." *Transactions of the American Fisheries Society* 82 (1952): 178–192.

———. "Angler's Arithmetic." In *Fish and Wildlife: A Memorial to W. J. K. Harkness*, ed. John R. Dymond, 55–71. Toronto: Longmans, 1964.

Fry, Fred E., and Kenneth E. F. Watt. "Yields of Year Classes of the Smallmouth Bass Hatched in the Decade of 1940 in Manitoulin Island Waters." *Transactions of the American Fisheries Society* 85 (1957): 136–143.

Galison, Peter. *How Experiments End*. Chicago: University of Chicago Press, 1987.

Galison, Peter, and Bruce Hevly, eds. *Big Science: The Growth of Large-Scale Research*. Stanford: Stanford University Press, 1992.

Gelpe, Marcia R., and A. Dan Tarlock. "The Uses of Scientific Information in Environmental Decisionmaking." *Southern California Law Review* 48 (1974): 371–427.

Gillette, Robert. "Reactor Emissions: AEC Guidelines Move Toward Critics Position." *Science* 172 (1971): 1215–1216.

———. "Nuclear Safety I: The Roots of Dissent." *Science* 177 (1972): 771–776.

———. "Nuclear Safety II: The Years of Delay." *Science* 177 (1972): 867–871.

Girard, Michel F. "The Commission of Conservation as a Forerunner to the National Research Council, 1909–1921." *Scientia Canadensis* 15, no. 2 (1991): 19–40.

Godwin, Harry. "The Sub-Climax and Deflected Succession." *Journal of Ecology* 17 (1929): 144–147.

———. "The 'Sedge' and 'Litter' of Wicken Fen." *Journal of Ecology* 17 (1929): 148–160.

———. *Cambridge and Clare*. Cambridge: Cambridge University Press, 1985.

Gofman, John W., and Arthur R. Tamplin. *Poisoned Power*. Emmaus, Pa.: Rodale, 1971.

Goldstein, Robert A., and W. F. Harris. "SERENDIPITY—A Watershed-Level Simulation Model of Forest Biomass Dynamics." EDFB Memo Report 72–168, 1972.

Golley, Frank B. *A History of the Ecosystem Concept in Ecology: More Than the Sum of the Parts*. New Haven: Yale University Press, 1993.

Goodwin, Arthur S. "Salt Water Trawls Tried in Lake Erie." *Canadian Fisherman*, March 1960, 29–31.

Gorham, Eville, Peter M. Vitousek, and William A. Reiners. "The Regulation of Chemical Budgets over the Course of Terrestrial Ecosystem Succession." *Annual Review of Ecology and Systematics* 10 (1979): 53–84.

Gosz, James R., Gene E. Likens, and F. Herbert Bormann. "Nutrient Content of Litter Fall on the Hubbard Brook Experimental Forest, New Hampshire." *Ecology* 53 (1972): 769–784.

———. "Nutrient Release from Decomposing Leaf and Branch Litter in the Hubbard Brook Forest, New Hampshire." *Ecological Monographs* 43 (1973): 173–191.

———. "Organic Matter and Nutrient Dynamics of the Forest Floor in the Hubbard Brook Forest." *Oecologia* 22 (1976): 305–320.

Gottlieb, Robert. *Forcing the Spring: The Transformation of the American Environmental Movement*. Washington: Island Press, 1993.

Graham, Edward H. *Natural Principles of Land Use*. London: Oxford University Press, 1944.

Green, Bryn. *Countryside Conservation: The Protection and Management of Amenity Ecosystems*. London: G. Allen and Unwin, 1981.

Greenberg, Daniel. *The Politics of Pure Science*. New York: New American Library, 1967.

Greenwood, Ted. *Knowledge and Discretion in Government Regulation*. New York: Praeger, 1984.

Gulland, John A. "Fishery Management: New Strategies for New Conditions." *Transactions of the American Fisheries Society* 107 (1978): 1–11.

Gummett, Philip J., and Geoffrey L. Price. "An Approach to the Central Planning of British Science: The Formation of the Advisory Council on Scientific Policy." *Minerva* 15 (1977): 119–143.

Haas, Peter M. *Saving the Mediterranean: The Politics of International Environmental Cooperation*. New York: Columbia University Press, 1990.

Hacker, Barton C. *The Dragon's Tail: Radiation Safety in the Manhattan Project, 1942–1946*. Berkeley: University of California Press, 1987.

Hagen, Joel B. *An Entangled Bank: The Origins of Ecosystem Ecology*. New Brunswick, N.J.: Rutgers University Press, 1992.

Hall, Charles A. S. "Models and the Decision Making Process: The Hudson River Power Plant Case." In *Ecosystem Modeling in Theory and Practice,* ed. Charles A. S. Hall and John W. Day, 345–364. New York: John Wiley, 1977.

Hall, Ronald J., Gene E. Likens, Sandy B. Fiance, and George R. Hendry. "Experimental Acidification of a Stream in the Hubbard Brook Experimental Forest, New Hampshire." *Ecology* 61 (1980): 976–989.

Harkness, W. J. K. "Ontario Fisheries Research Laboratory." 1929. Typescript. Ontario Government Archives, Toronto, File RG 1–289-1-2, Advisory Committee for Harkness Laboratory.

Harkness, W. J. K., and Fred E. Fry. "Game Fish Management in Algonquin Park Lakes." *North American Wildlife Conference* 7 (1942): 398–404.

Harris, W. F., and David E. Reichle. "Evaluating Forest Productivity in an Ecosystem Context." *Proceedings of the Agricultural Research Institute* (1973): 57–68.

Harvey, Harold H. "Aquatic Environmental Quality: Problems and Proposals." *Journal of the Fisheries Research Board of Canada* 33 (1976): 2634–2670.

Harvey, Harold H., and J. F. Coombs. "Physical and Chemical Limnology of the Lakes of Manitoulin Island." *Journal of the Fisheries Research Board of Canada* 28 (1971): 1883–1897.

Harvey, Harold H., and Glen H. Geen. "Introduction to Limnology in Canada." *Journal of the Fisheries Research Board of Canada* 31 (1974): 501–504.

Harvey, H. John. "Changing Attitudes to Nature Conservation: The National Trust." *Biological Journal of the Linnean Society* 32 (1987): 149–159.

Harwood, Jonathan. *Styles of Scientific Thought: The German Genetics Community, 1900–1933*. Chicago: University of Chicago Press, 1993.

Hasler, Arthur D. "Experimental Limnology." *BioScience* 14, no. 7 (1964): 36–38.

Hays, Samuel P. *Beauty, Health, and Permanence: Environmental Politics in the United States, 1955–1985*. Cambridge: Cambridge University Press, 1987.

Heidenheimer, Arnold J., Hugh Heclo, and Carolyn Teich Adams. "Environmental Policy." In *Comparative Public Policy: The Politics of Social Choice in America, Europe, and Japan,* ed. Arnold J. Heidenheimer, Hugh Heclo, and Carolyn Teich Adams, 308–344. New York: St. Martin's, 1990.

Henderson, H. F., Richard A. Ryder, and A. W. Kudhongania. "Assessing Fishery Potentials of Lakes and Reservoirs." *Journal of the Fisheries Research Board of Canada* 30 (1973): 2000–2009.

Hett, J. M. "Land-Use Changes in East Tennessee and a Simulation Model Which Describes These Changes for Three Counties." ORNL-IBP Report 71–8, 1971.

Hilborn, Ray. "Living with Uncertainty in Resource Management." *North American Journal of Fisheries Management* 7 (1987): 1–5.

Holifield, Chet. "Remarks Concerning Nuclear Power Reactor Licensing Problems." In Joint Committee for Atomic Energy, *AEC Licensing Procedure and Related Legislation*. Washington, D.C.: Government Printing Office, 1971.

Hollaender, Alexander. "The Biology Division of Oak Ridge National Laboratory." *AIBS Bulletin* 7 (1957): 10–14.

Holmes, Richard T., R. E. Bonney, Jr., and S. W. Pacala. "Guild Structure of the Hubbard Brook Bird Community: A Multivariate Approach." *Ecology* 60 (1979): 512–520.

Holmes, Richard T., T. W. Sherry, and S. E. Bennett. "Diurnal and Individual Variability in the Foraging Behavior of American Redstarts (*Setophaga ruticilla*)." *Oecologia* 36 (1978): 141–149.

Holmes, Richard T., and Frank W. Sturges. "Annual Energy Expenditure by the Avifauna of a Northern Hardwoods Ecosystem." *Oikos* 24 (1973): 24–29.

Hooper, Frank F. "Applications of Ecology to Environmental Assessment: The Role of Ecologists in the Decision-Making Process." In *Environmental Sciences Laboratory Dedication, February 26 and 27, 1979*, ed. Stanley I. Auerbach and N. T. Millemann, 45–51. Oak Ridge National Laboratory, ORNL-5700, 1980.

Hoover, M. D. "Effect of Removal of Forest Vegetation Upon Water Yields." *Transactions of the American Geophysical Union,* part 6 (1944): 969–975.

Hornbeck, James W., and Gene E. Likens. "The Ecosystem Concept for Determining the Importance of Chemical Composition of Snow." In *Advanced Concepts and Techniques in the Study of Snow and Ice Resources,* comp. Henry S. Santeford and James L. Smith, 139–151. Washington, D.C.: National Academy of Science, 1974.

Hornbeck, James W., Gene E. Likens, and John S. Eaton. "Seasonal Patterns in Acidity of Precipitation and Their Implications for Forest Stream Ecosystems." *Water, Air, and Soil Pollution* 7 (1977): 355–365.

Hornbeck, James W., Robert S. Pierce, and C. Anthony Federer. "Streamflow Changes After Forest Clearing in New England." *Water Resources Research* 6 (1970): 1124–1132.

Hurst, J. Willard. *Law and Social Order in the United States*. Ithaca: Cornell University Press, 1977.

Hutchinson, G. Evelyn. "Bio-Ecology." *Ecology* 21 (1940): 267–268.

———. "Circular Causal Systems in Ecology." *Annals of the New York Academy of Science* 40 (1948): 221–246.

Hutchinson, G. Evelyn, and A. Wollack. "Studies on Connecticut Lake Sediments, II. Chemical Analysis of a Core from Linsley Pond, North Brantford." *American Journal of Science* 238 (1940): 483–517.

Huxley, Julian S. *TVA: Adventure in Planning*. London: London Architectural Press, 1946.

————. *Memories*. Vol. 2. London: George Allen and Unwin, 1973.

Hywel-Davies, Jeremy, Valerie Thom, and Linda Bennett. *The Macmillan Guide to Britain's Nature Reserves*. London: Macmillan, 1986.

Jasanoff, Sheila. "The Problem of Rationality in American Health and Safety Regulation." In *Expert Evidence: Interpreting Science in the Law*, ed. R. Smith and B. Wynne, 151–183. New York: Routledge, 1989.

————. "Science, Politics, and the Renegotiation of Expertise at EPA." *Osiris*, 2nd ser., 7 (1992): 195–217.

Jeffrey, Michael. "Science and the Tribunal: Dealing with Scientific Evidence in the Adversarial Arena." *Alternatives* 15, no. 2 (1988): 24–30.

Johnson, Noye M., Gene E. Likens, F. Herbert Bormann, D. W. Fisher, and Robert S. Pierce. "A Working Model for the Variation in Stream Water Chemistry at the Hubbard Brook Experimental Forest, New Hampshire." *Water Resources Research* 5 (1969): 1353–1363.

Johnson, Noye M., Gene E. Likens, F. Herbert Bormann, and Robert S. Pierce. "Rate of Chemical Weathering of Silicate Minerals in New Hampshire." *Geochimica et Cosmochimica Acta* 32 (1968): 531–545.

Johnstone, Kenneth. *The Aquatic Explorers: A History of the Fisheries Research Board of Canada*. Toronto: University of Toronto Press–Fisheries Research Board of Canada, 1977.

Kaye, Stephen V., and Sydney J. Ball. "Systems Analysis of a Coupled Compartment Model for Radionuclide Transfer in a Tropical Environment." In *Symposium on Radioecology, Proceedings of the Second National Symposium, Ann Arbor, Michigan, 1967*, ed. Daniel J. Nelson and F. C. Evans, 731–739. Washington, D.C.: Atomic Energy Commission, 1969.

Kaye, Stephen V., and Paul B. Dunaway. "Bioaccumulation of Radioactive Isotopes by Herbivorous Small Mammals." *Health Physics* 7 (1962): 205–217.

————. "Estimation of Dose Rate and Equilibrium State from Bioaccumulation of Radionuclides by Mammals." In *Radioecology: Proceedings of the First National Symposium on Radioecology, September 10–15, 1961*, ed. Vincent Schultz and Alfred W. Klement, 107–111. New York: Reinhold, 1963.

Keating, William T. "Politics, Energy, and the Environment: The Role of Technology Assessment." *American Behavioral Scientist* 19 (1975): 37–74.

Kevles, Daniel J. *The Physicists: The History of a Scientific Community in Modern America*. Cambridge: Harvard University Press, 1987.

Kimball, Marcus. "Nature Conservancy (Research), 21 July 1961." *Parliamentary Debates*, Commons, 1706.

Kingsland, Sharon. "An Elusive Science: Ecological Enterprise in the Southwestern United States." In *Science and Nature: Essays in the History of the Environmental Sciences*, ed. Michael Shortland, 151–179. Oxford: British Society for the History of Science, 1993.

Kittredge, J. *Forest Influences*. New York: McGraw-Hill, 1948.

Kohler, Robert E. *From Medical Chemistry to BioChemistry: The Making of a Biomedical Discipline*. Cambridge: Cambridge University Press, 1982.

Krumholz, Louis A. "A Summary of Findings of the Ecological Survey of White Oak Creek, Roane County, Tennessee, 1950–1953." U.S. AEC Report ORO-132, 1954.

Kwa, Chunglin. "Representations of Nature Mediating Between Ecology and Science Policy: The Case of the International Biological Programme." *Social Studies of Science* 17 (1987): 413–442.

———. "Modeling the Grasslands." *Historical Studies in the Physical and Biological Sciences* 24 (1993): 125–155.

———. "Radiation Ecology, Systems Ecology and the Management of the Environment." In *Science and Nature: Essays in the History of the Environmental Sciences,* ed. Michael Shortland, 213–241. British Society for the History of Science, 1993.

Laboratory for Experimental Limnology. "Report of the Laboratory for Experimental Limnology for the Years 1952 and 1953." *Ontario Department of Lands and Forests, Research Report,* no. 28, 1954.

———. "The Laboratory for Experimental Limnology, 1954–1959." *Ontario Department of Lands and Forests, Research Report,* no. 44, 1961.

Lakoff, Sanford A. "Congress and National Science Policy." *Political Science Quarterly* 89 (1974): 589–611.

Lambert, Richard S., and Paul Pross. *Renewing Nature's Wealth: A Centennial History of the Public Management of Lands, Forests and Wildlife in Ontario, 1763–1967.* Toronto: Ontario Department of Lands and Forests, 1967.

Langford, R. R. "Fisheries Research in Algonquin Park." *Sylva* 4, no. 2 (1948): 25–32.

———. "Fertilization of Lakes in Algonquin Park, Ontario." *Transactions of the American Fisheries Society* 78 (1950): 133–144.

———. "Opeongo Laboratory." *Sylva* 8 (1952): 25–28.

Lapworth, E. D. "The Effect of Fry Plantings on Whitefish Production in Eastern Lake Ontario." *Journal of the Fisheries Research Board of Canada* 13 (1956): 547–558.

Larkin, Peter A. "A Confidential Memorandum on Fisheries Science." In *World Fisheries Policy: Multidisciplinary Views,* ed. B.J. Rothschild, 189–197. Seattle: University of Washington Press, 1972.

———. "An Epitaph for the Concept of Maximum Sustained Yield." *Transactions of the American Fisheries Society* 106 (1977): 1–11.

Larkin, Peter A., and Norman J. Wilimovsky. "Contemporary Methods and Future Trends in Fishery Management and Development." *Journal of the Fisheries Research Board of Canada* 30 (1973): 1948–1957.

Laurie, M. V. "The Present Productivity of Forests." In *The Biological Productivity of Britain,* ed. William Yapp and W. Brunsdon, 73–82. London: Institute of Biology, 1958.

Lawler, G. H. "Fluctuations in the Success of Year-Classes of Whitefish Populations with Special Reference to Lake Erie." *Journal of the Fisheries Research Board of Canada* 22 (1965): 1197–1227.

Leiss, William, and Christina Chociolko. *Risk and Responsibility*. Montreal: McGill–Queen's University Press, 1994.

Leopold, Aldo. *Game Management*. New York: Scribner, 1933.

Leslie, A. P. "DDT in Ontario's Forests." *Canadian Geographical Journal*, October 1945, 186–197.

Levins, Richard. "The Strategy of Model Building in Population Biology." *American Scientist* 54 (1966): 421–431.

Likens, Gene. "An Unusual Distribution of Àlgae in an Antarctic Lake." *Bulletin of the Torrey Botanical Club* 91 (1964): 213–217.

———. "Final Report to the National Science Foundation, Undergraduate Research Participation Program, 1966–67." October 16, 1967. Hubbard Brook Ecosystem Study Files, Institute of Ecosystem Studies.

———. "A Priority for Ecological Research." *Bulletin of the Ecological Society of America* 64 (1983): 234–243.

Likens, Gene E., and F. Herbert Bormann. "Effects of Forest Clearing on the Northern Hardwood Forest Ecosystem and Its Biogeochemistry." In *Proceedings of the First International Congress of Ecology*, 330–335. The Hague: Center for Agricultural Publishing and Documentation, 1974.

———. "Hubbard Brook Ecosystem Study, Response to NSF Questions." 22 February 1976.

Likens, Gene E., F. Herbert Bormann, and Noye M. Johnson. "Nitrification: Importance to Nutrient Losses from a Cutover Forested Ecosystem." *Science* 163 (1969): 1205–1206.

Likens, Gene E., F. Herbert Bormann, Noye M. Johnson, D. W. Fisher, and Robert S. Pierce. "Effects of Forest Cutting and Herbicide Treatment on Nutrient Budgets in the Hubbard Brook Watershed-Ecosystem." *Ecological Monographs* 40 (1970): 23–47.

Likens, Gene E., F. Herbert Bormann, Noye M. Johnson, and Robert S. Pierce. "The Calcium, Magnesium, Potassium, and Sodium Budgets for a Small Forested Ecosystem." *Ecology* 48 (1967): 772–785.

Likens, Gene E., F. Herbert Bormann, Robert S. Pierce, John S. Eaton, and Noye M. Johnson. *Biogeochemistry of a Forested Ecosystem*. New York: Springer-Verlag, 1977.

Likens, Gene E., F. Herbert Bormann, Robert S. Pierce, and W. A. Reiners. "Recovery of a Deforested Ecosystem." *Science* 199 (1978): 492–496.

Likens, Gene E., and Arthur D. Hasler. "Movements of Radiosodium (Na^{24}) Within an Ice-Covered Lake." *Limnology and Oceanography* 7 (1962): 48–56.

Likens, Gene E., ed. *An Ecosystem Approach to Aquatic Ecology: Mirror Lake and Its Environment*. New York: Springer-Verlag, 1985.

Likens, Phyllis C. *Publications of the Hubbard Brook Ecosystem Study*. Millbrook, N.Y.: Institute of Ecosystem Studies, 1988.

Lilienfeld, Robert. *The Rise of Systems Theory: An Ideological Analysis*. New York: John Wiley and Sons, 1978.

Lindeman, Raymond L. "The Trophic-Dynamic Aspect of Ecology." *Ecology* 23 (1942): 319–418.

Lindop, Patricia J., and J. Rotblat. "Radiation Pollution of the Environment." *Bulletin of the Atomic Scientists* 27 (1971): 17–24.

Lipschutz, Ronnie D. *Radioactive Waste: Politics, Technology, and Risk*. Cambridge: Ballinger, 1980.

Loftus, Kenneth H. "A New Approach to Fisheries Management and F. E. J. Fry's Role in Its Development." *Journal of the Fisheries Research Board of Canada* 33 (1976): 321–325.

Loftus, Kenneth H., M. G. Johnson, and Henry A. Regier. "Federal-Provincial Strategic Planning for Ontario Fisheries: Management Planning for the 1980s." *Journal of the Fisheries Research Board of Canada* 35 (1978): 916–927.

Loftus, Kenneth H., and Henry A. Regier, eds. "Proceedings of the Symposium on Salmonid Communities in Oligotrophic Lakes (SCOL)." *Journal of the Fisheries Research Board of Canada* 29 (1972): 611–986.

Loucks, Orie L. "Systems Methods in Environmental Court Actions." In *Systems Analysis and Simulation in Ecology*, ed. Bernard C. Patten, 2: 419–473. New York: Academic Press, 1972.

Loucks, Orie L. "The Role of Basic Ecological Knowledge in the Mitigation of Impacts from Complex Technological Systems: Agriculture, Transportation, and Urban." In *Preserving Ecological Systems: The Agenda for Long-Term Research and Development*, ed. Sidney Draggan, John J. Cohrssen, and Richard Morrison, 71–92. New York: Praeger, 1987.

Lovins, Amory. *Soft Energy Paths: Toward a Durable Peace*. Cambridge, Mass.: Friends of the Earth–Ballinger, 1977.

Lowe, Philip D. "Values and Institutions in the History of British Nature Conservation." In *Conservation in Perspective*, ed. A. Warren and F. B. Goldsmith, 329–352. Chichester: John Wiley and Sons, 1983.

Lowe, Philip D., and Jane Goyder. *Environmental Groups in Politics*. London: George Allen and Unwin, 1983.

Lowe, Philip, Graham Cox, Malcolm MacEwen, Tim O'Riordan, and Michael Winter. *Countryside Conflicts: The Politics of Farming, Forestry and Conservation*. Aldershot, Hants, England: Gower/Maurice Temple Smith, 1986.

Lowrance, William W. *Of Acceptable Risk: Science and the Determination of Safety*. Los Altos, Calif.: William Kaufmann, 1976.

Lubchenco, Jane, et al. "The Sustainable Biosphere Initiative: An Ecological Research Agenda." *Ecology* 72 (1991): 371–412.

McEvoy, Arthur F. *The Fisherman's Problem: Ecology and Law in the California Fisheries, 1850–1980*. Cambridge: Cambridge University Press, 1986.

McGucken, William. "On Freedom and Planning in Science: The Society for Freedom in Science, 1940–46." *Minerva* 16 (1978): 42–72.

———. *Scientists, Society, and State: The Social Relations of Science Movement in Great Britain, 1931–1947*. Columbus: Ohio State University Press, 1984.

McIntosh, Robert P. *The Background of Ecology: Concept and Theory*. Cambridge: Cambridge University Press, 1985.

Mann, Dean E. *Environmental Policy Implementation: Planning and Management Options and Their Consequences*. Lexington, Mass.: Lexington Books, 1982.

Marcus, Alfred Allen. *Promise and Performance: Choosing and Implementing an Environmental Policy.* Westport, Conn.: Greenwood, 1980.

Margalef, Ramon. "Information Theory in Ecology." *General Systems* 3 (1958): 36–71.

Marks, Peter L. "The Role of Pin Cherry (*Prunus pensylvanica* L.) in the Maintenance of Stability in Northern Hardwood Ecosystems." *Ecological Monographs* 44 (1974): 73–88.

Marks, Peter L., and F. Herbert Bormann. "Revegetation Following Forest Cutting: Mechanisms for Return to Steady-State Nutrient Cycling." *Science* 176 (1972): 914–915.

Marsh, George P. *Report, Made Under Authority of the Legislature of Vermont, on the Artificial Propagation of Fish.* Burlington, Vt.: Free Press Print, 1857.

Martin, C. Wayne. "Precipitation and Streamwater Chemistry in an Undisturbed Forested Watershed in New Hampshire." *Ecology* 60 (1979): 36–42.

Marx, Leo. "Environmental Degradation and the Ambiguous Social Role of Science and Technology." *Journal of the History of Biology* 25 (1992): 449–468.

May, Robert M. *Stability and Complexity in Model Ecosystems.* Princeton: Princeton University Press, 1973.

Mazuzan, George T., and J. Samuel Walker. *Controlling the Atom: The Beginnings of Nuclear Regulation, 1946–1962.* Berkeley: University of California Press, 1984.

McGucken, William. "The Canadian Federal Government, Cultural Eutrophication, and the Regulation of Detergent Phospates, 1970." *Environmental Review* 13 (1989): 155–166.

Metzger, H. Peter. *The Atomic Establishment.* New York: Simon and Schuster, 1972.

Meyer, Judy L. "Dynamics of Phosphorus and Organic Matter During Leaf Decomposition in a Forest Stream." *Oikos* 34 (1980): 44–53.

Mitman, Gregg. *The State of Nature: Ecology, Community, and American Social Thought, 1900–1950.* Chicago: University of Chicago Press, 1992.

Moore, Norman W. *The Bird of Time: The Science and Politics of Nature Conservation—A Personal Account.* Cambridge: Cambridge University Press, 1987.

Moravcsik, Michael J. "Reflections on National Laboratories." *Bulletin of the Atomic Scientists,* February 1970, 11–16.

Morgan, Karl Z. "Foreword." *Health Physics* 1 (1958): 1–2.

Morgan, Karl Z., and Stanley I. Auerbach. "Need for Reserving Melton Valley for Long-Range Ecological Studies." ORNL/CF-57/12/25, 1957.

Morison, Robert S. "Problems of Science, Goals and Priorities." In *Science and the Evolution of Public Policy,* ed. James A. Shannon, 47–66. New York: Rockefeller University Press, 1973.

Morris, S. "Will the Great Lakes Be Fished Out?" *Canadian Fisherman,* September 1958, 6–11.

———. "Discouraging Year for the Great Lakes." *Canadian Fisherman,* June 1960, 28.

Morrison, Herbert. "Planned Research." *Advancement of Science* 3 (1946): 296–299.

———. *Government and Parliament: A Survey from the Inside.* London: Oxford University Press, 1954.

Morton, Roy J., and Edward G. Struxness. "Ground Disposal of Radioactive Wastes." *American Journal of Public Health* 46 (1956): 156–163.

Naess, Arne. "The Shallow and the Deep, Long-Range Ecology Movement: A Summary." *Inquiry* 16 (1973): 95–100.

National Academy of Sciences. *Federal Support of Basic Research in Institutions of Higher Learning: Historical Review.* Washington, D.C.: National Research Council, 1964.

"Nature Preservation in England." *Nature* 157 (1946): 277–278.

Neel, R. B., and Jerry S. Olson. "Use of Analog Computers for Simulating the Movement of Isotopes in Ecological Systems." Oak Ridge National Laboratory, ORNL-3172, January 1962.

Nelkin, Dorothy. *Nuclear Power and Its Critics: The Cayuga Lake Controversy.* Ithaca: Cornell University Press, 1971.

———. "Scientists and Professional Responsibility: The Experience of American Ecologists." *Social Studies of Science* 7 (1977): 75–95.

———. "Science, Technology, and Political Conflict: Analyzing the Issues." In *Controversy: The Politics of Technical Decisions,* 3rd ed., ed. Dorothy Nelkin, ix–xxv. Newbury Park, Calif.: Sage, 1992.

Nelson, Daniel J., and Donald C. Scott. "Role of Detritus in the Productivity of a Rock-Outcrop Community in a Piedmont Stream." *Limnology and Oceanography* 7 (1962): 396–413.

Nicholson, E. Max. *Birds in England: An Account of the State of Our Bird Life and a Criticism of Bird Protection.* London: Chapman and Hall, 1926.

———. "Ecological Research and Farming." *Advancement of Science* 13 (1957): 439–443.

———. *The System: The Misgovernment of Modern Britain.* London: Hodder and Stoughton, 1967.

———. *The Environmental Revolution: A Guide for the New Masters of the World.* London: Hodder and Stoughton, 1970.

———. *The New Environmental Age.* Cambridge: Cambridge University Press, 1987.

Oak Ridge National Laboratory. *Health Physics Division Progress Reports* (1955–1969).

———. *Ecological Sciences Division Annual Progress Reports* (1970–1972).

———. *Environmental Sciences Division Annual Progress Reports* (1973–1980).

Odum, Eugene P. *Fundamentals of Ecology.* Philadelphia: Saunders, 1953.

———. "Consideration of the Total Environment in Power Reactor Waste Disposal." In *Proceedings of the First International Conference on Peaceful Uses of Atomic Energy* 13 (1956): 350–353.

———. "Panel Discussion on Education and Research Training." In *Radioecology: Proceedings of the First National Symposium on Radioecology, September 10–15, 1961,* ed. Vincent Schultz and Alfred W. Klement, 643–645. New York: Reinhold, 1963.

———. "Feedback Between Radiation Ecology and General Ecology." *Health Physics* 11 (1965): 1257–1262.

————. "Ecosystem Theory in Relation to Man." In *Ecosystem Structure and Function,* ed. John A. Wiens, 11–24. Corvallis: Oregon State University Press, 1972.

————. "The Emergence of Ecology as a New Integrative Discipline." *Science* 195 (1977): 1289–1293.

Odum, Howard T. "The Stability of the World Strontium Cycle." *Science* 114 (1951): 407–411.

————. "Trophic Structure and Productivity of Silver Springs, Florida." *Ecological Monographs* 27 (1957): 55–112.

Odum, Howard T., and R. C. Pinkerton. "Time's Speed Regulator: The Optimum Efficiency for Maximum Power Output in Physical and Biological Systems." *American Scientist* 43 (1955): 331–343.

Olson, Jerry S. "Rates of Succession and Soil Changes on Southern Lake Michigan Sand Dunes." *Botanical Gazette* 119 (1958): 125–170.

————. "Analog Computer Models for Movement of Nuclides Through Ecosystems." In *Radioecology: Proceedings of the First National Symposium on Radioecology, September 10–15, 1961,* ed. Vincent Schultz and Alfred W. Klement, 121–125. New York: Reinhold, 1963.

————. "Energy Storage and the Balance of Producers and Decomposers in Ecological Systems." *Ecology* 44 (1963): 322–331.

————. "Equations for Cesium Transfer in a *Liriodendron* Forest." *Health Physics* 11 (1965): 1385–1392.

————. "Carbon Cycles and Temperate Woodlands." In *Analysis of Temperate Forest Ecosystems,* ed. David E. Reichle, 226–241. Berlin: Springer Verlag, 1970.

O'Neill, Robert V. "Pathway Analysis: A Preliminary Application of Systems Ecology to Nuclear Facility Safety Evaluation." ORNL/CF-70/3/25, 1970.

————. "Tracer Kinetics in Total Ecosystems: A Systems Analysis Approach." In *Nuclear Techniques in Environmental Pollution,* Proceedings, International Atomic Energy Agency Symposium 142a/45, 693–705. Vienna, 1971.

————. "Modeling in the Eastern Deciduous Forest Biome." In *Systems Analysis and Simulation in Ecology,* vol. 3, ed. Bernard C. Patten, 49–72. New York: Academic Press, 1975.

O'Neill, Robert V., B. S. Ausmus, D. R. Jackson, R. I. Van Hook, P. Van Voris, C. Washburne, and A. P. Watson. "Monitoring Terrestrial Ecosystems by Analysis of Nutrient Export." *Water, Air and Soil Pollution* 8 (1977): 271–277.

O'Neill, Robert V., and O. W. Burke. "A Simple Systems Model for DDT and DDE Movement in the Human Food-Chain." ORNL-IBP Report 71–9, 1971.

O'Neill, Robert V., Robert A. Goldstein, Hank H. Shugart, and J. B. Mankin. "Terrestrial Ecosystem Energy Model." EDFB Memo Report 72–19, 1972.

O'Neill, Robert V., W. F. Harris, B. S. Ausmus, and David E. Reichle. "A Theoretical Basis for Ecosystem Analysis with Particular Reference to Element Cycling." In *Mineral Cycling in Southeastern Ecosystems,* ed. F. G. Howell et al., 28–40. Energy Research and Development Administration, ERDA Symposium Series, CONF—740513, 1975.

Ontario Department of Lands and Forests. *Annual Reports* (1948–1971).

"Open-Air Laboratories: The Art of Nature Conservancy in Britain." *Times* (London), 21 August 1954.

O'Riordan, Timothy. *Environmentalism*, 2nd ed. London: Pion, 1981.

O'Sullivan, P. E. "The Ecosystem-Watershed Concept in the Environmental Sciences—A Review." *International Journal of Environmental Studies* 13 (1979): 273–281.

Ott, W. R., ed. *Proceedings of the Conference on Environmental Modeling and Simulation*. U.S. Environmental Protection Agency, Office of Research and Development, 1976.

Ovington, J. Derek. "The Afforestation of the Culbin Sands." *Journal of Ecology* 38 (1950): 303–319.

———. "The Afforestation of the Tentsmuir Sands." *Journal of Ecology* 39 (1951): 363–375.

———. "A Study of Invasion by *Holcus mollis* L." *Journal of Ecology* 41 (1953): 35–52.

———. "Studies of the Development of Woodland Conditions under Different Trees. I. Soils pH." *Journal of Ecology* 41 (1953): 13–34.

———. "A Comparison of Rainfall in Different Woodlands." *Journal of Forestry* 27 (1954): 41–53.

———. "Studies of the Development of Woodland Conditions Under Different Trees II. The Forest Floor." *Journal of Ecology* 42 (1954): 71–80.

———. "The Form, Weights and Productivity of Tree Species Grown in Close Stands." *New Phytologist* 55 (1956): 289–304.

———. "The Volatile Matter, Organic Carbon and Nitrogen Contents of Tree Species Grown in Close Stands." *New Phytologist* 56 (1957): 1–11.

———. "Dry-Matter Production by *Pinus sylvestris* L." *Annals of Botany* 21 (1957): 287–314.

———. "Some Biological Considerations of Forest Production." In *The Biological Productivity of Britain*, ed. William B. Yapp and D. J. Watson, 83–90. London: Institute of Biology, 1958.

———. "Some Aspects of Energy Flow in Plantations of *Pinus sylvestris* L." *Annals of Botany* 25 (1961): 12–20.

———. "Quantitative Ecology and the Woodland Ecosystem Concept." *Advances in Ecological Research* 1 (1962): 103–192.

———. *Woodlands*. London: English Universities Press, 1965.

Ovington, J. Derek, and D. Heitkamp. "The Accumulation of Energy in Forest Plantations in Britain." *Journal of Ecology* 48 (1960): 639–646.

Ovington, J. Derek, and D. B. Lawrence. "Strontium 90 in Maize Field, Cattail Marsh and Oakwood Ecosystems." *Journal of Applied Ecology* 1 (1964): 175–181.

Ovington, J. Derek, and H. A. I. Madgwick. "The Growth and Composition of Natural Stands of Birch 1. Dry-Matter Production." *Plant and Soil* 10 (1959): 389–400.

Ozawa, Connie P. *Recasting Science: Consensual Procedures in Public Policy Making*. Boulder, Colo.: Westview, 1991.

Paelhke, Robert, and Doug Torgerson, eds. *Managing Leviathan: Environmental Politics and the Administrative State*. Peterborough, Ontario: Broadview, 1990.

Palmer, David D., Lauriston R. King, and W. Wayne Shannon. "Oceanographers and the US Federal Patron: Perceptions of Agency-University Relations." *Social Studies of Science* 18 (1988): 73–90.

Patten, Bernard C. "An Introduction to the Cybernetics of the Ecosystem: The Trophic-Dynamic Aspect." *Ecology* 40 (1959): 221–231.

———. "Competitive Exclusion." *Science* 134 (1961): 1599–1601.

———. "The Systems Approach in Radiation Ecology." Oak Ridge National Laboratory, ORNL-TM-1008, November 16, 1964.

———. "Community Organization and Energy Relationships in Plankton." Oak Ridge National Laboratory, ORNL-3634, 1965.

———. "Systems Ecology: A Course Sequence in Mathematical Ecology." *BioScience* 16 (1966): 593–599.

Patten, Bernard C., and Martin Witkamp. "Systems Analysis of 134 Cesium Kinetics in Terrestrial Microcosms." *Ecology* 48 (1967): 813–824.

Pauly, Philip J. "The Appearance of Academic Biology in Late Nineteenth-Century America." *Journal of the History of Biology* 17 (1984): 369–397.

Pearsall, William H. "The Soil Complex in Relation to Plant Communities." *Journal of Ecology* 26 (1938): 180–193, 194–205.

———. "A Milestone in Plant Ecology." *Journal of Ecology* 28 (1940): 241–244.

———. *Mountains and Moorlands*. London: Collins, 1950.

———. "Habitat, Vegetation and Man." *Nature* 171 (1953): 367–368.

———. "Growth and Production." *Advancement of Science* 11 (1954): 232–241.

———. "Production Ecology." *Science Progress* 47 (1959): 106–111.

Petulla, Joseph M. *American Environmentalism: Values, Tactics, Priorities*. College Station: Texas A & M University Press, 1980.

Phillipson, Donald J. C. "The National Research Council of Canada: Its Historiography, Its Chronology, Its Bibliography." In *Building Canadian Science: The Role of the National Research Council*, ed. Richard A. Jarrell and Yves Gingras, 177–200. Ottawa: Canadian Science and Technology Historical Association, 1992.

Pierce, Robert S., C. Wayne Martin, and Gene E. Likens. "Clearcutting and Nutrient Loss in New Hampshire." Paper delivered before the Society of American Foresters, Section Meeting, Boston, 8 March 1972.

Pimlott, Douglas H. "Scientific Activities in Fisheries." In *Scientific Activities in Fisheries and Wildlife Resources*, ed. Douglas H. Pimlott, C. James Kerswill, and John Bider, 87–110. Ottawa: Information Canada, 1971.

Pinder, John, ed. *Fifty Years of Political and Economic Planning: Looking Forward, 1931–1981*. London: Heinemann, 1981.

Poole, J. B., and Kay Andrews, eds. *The Government of Science in Britain*. London: Weidenfeld and Nicolson, 1972.

Poore, Duncan M. E. "Changing Attitudes in Nature Conservation: The Nature Conservancy and Nature Conservancy Council." *Biological Journal of the Linnean Society* 32 (1987): 179–187.

President's Science Advisory Committee, Environmental Pollution Panel. *Restoring *the Quality of Our Environment.* Washington, D.C.: Government Printing Office, 1965.

Primack, Joel, and Frank von Hippel. *Advice and Dissent: Scientists in the Political Arena.* New York: Basic Books, 1974.

Pritchard, Donald W., and Arnold B. Joseph. "Disposal of Radioactive Wastes: Its History, Status, and Possible Impact on the Environment." In *Radioecology: Proceedings of the First National Symposium on Radioecology, September 10–15, 1961,* ed. Vincent Schultz and Alfred W. Klement, 27–31. New York: Reinhold, 1963.

Pye-Smith, Charlie, and Chris Rose. *Crisis and Conservation: Conflict in the British Countryside.* Harmondsworth: Penguin, 1984.

Rackham, Oliver. *The History of the Countryside.* London: J. M. Dent and Sons, 1986.

Ragotzkie, Robert A., and Gene E. Likens. "The Heat Balance of Two Antarctic Lakes." *Limnology and Oceanography* 9 (1964): 412–425.

Reagan, Michael. *Science and the Federal Patron.* New York: Oxford University Press, 1969.

Regier, Henry A. "Some Models of Walleye Production in Western Lake Erie." October 1962. Typescript.

———. "A Perspective on Research on the Dynamics of Fish Populations in the Great Lakes." *Progressive Fish-Culturalist* (1966): 3–18.

———. "Some Ecological Considerations on the Introduction of Non-Native Species into the Great Lakes." Paper presented at American Fisheries Society meeting, New Orleans, September 1969.

———. "Sequence of Exploitation of Stocks in Multispecies Fisheries in the Laurentian Great Lakes." *Journal of the Fisheries Research Board of Canada* 30 (1973): 1992–1999.

Regier, Henry A., Vernon C. Applegate, and Richard A. Ryder. "The Ecology and Management of the Walleye in Western Lake Erie." *Great Lakes Fishery Commission, Technical Report,* no. 15 (1969).

Regier, Henry A., and W. L. Hartman. "Lake Erie's Fish Community: 150 Years of Cultural Stresses." *Science* 180 (1973): 1248–1255.

Regier, Henry A., and H. Francis Henderson. "Towards a Broad Ecological Model of Fish Communities and Fisheries." *Transactions of the American Fisheries Society* 102 (1973): 56–72.

Regier, Henry A., and F. D. McCracken. "Science for Canada's Shelf-Seas Fisheries." *Journal of the Fisheries Research Board of Canada* 32 (1975): 1887–1932.

Reichle, David E. "Radioisotope Turnover and Energy Flow in Terrestrial Isopod Populations." *Ecology* 48 (1967): 351–366.

———. "Relation of Body Size to Food Intake, Oxygen Consumption, and Trace Element Metabolism in Forest Floor Arthopods." *Ecology* 49 (1968): 538–542.

———. "Advances in Ecosystem Analysis." *BioScience* 25 (1975): 257–264.

Reichle, David E., and Stanley I. Auerbach. "Analysis of Ecosystems." In *Challenging Biological Problems: Directions Towards Their Solutions,* ed. J. A. Behnke, 266–273. Arlington, Va.: American Institute of Biological Science, 1972.

Reichle, David E., and D. A. Crossley. "Investigation on Heterotrophic Productivity in Forest Insect Communities." In *Secondary Productivity of Terrestrial Ecosystems (Principles and Methods)*, ed. K. Petrusewicz, 563–587. Warsaw, 1967.

Reichle, David E., Paul B. Dunaway, and Daniel J. Nelson. "Turnover and Concentration of Radionuclides in Food Chains." *Nuclear Safety* 11 (1970): 43–55.

Reichle, David E., Robert V. O'Neill, Jerry S. Olson, and Lynda Kern. "Modeling Forest Ecosystems: Report of International Woodlands Workshop IBP/PT Section, August 14–26, 1972." EDFB-IBP-73-7, August 1973.

Richardson, Arthur H. *Conservation by the People: The History of the Conservation Movement in Ontario to 1970*. Toronto: University of Toronto Press, 1974.

Rigler, Frank H., "The Char Lake Project." In *Energy Flow—Its Biological Dimensions: A Summary of the IBP in Canada, 1964–1974*, ed. Thomas W. M. Cameron and L. W. Billingsley, 171–198. Ottawa: Royal Society of Canada, 1975.

Robertson, J. S. "Theory and Use of Tracers in Determining Transfer Rates in Biological Systems." *Physiological Reviews* 37 (1957): 133–154.

Robinson, Glen O. *The Forest Service*. Washington, D.C.: Resources for the Future, 1975.

Robinson, Roy. "Some Ecological Aspects of Afforestation and Forestry in Great Britain." *Forestry* 16 (1942): 1–12.

Ronayne, Jarlath. *Science in Government*. Caulfield: E. Arnold, 1984.

Rose, David J. "New Laboratories for Old." *Daedalus* 103, no. 3 (1974): 145–155.

Rose, Hilary, and Steven Rose. *Science and Society*. London: Allen Lane, the Penguin Press, 1969.

Rosenbaum, Walter A. *The Politics of Environmental Concern*. New York: Praeger, 1977.

Rosenberg, Charles. "Woods or Trees: Ideas and Actors in the History of Science." *Isis* 79 (1988): 565–570.

Royal Commission on the Future of the Toronto Waterfront. *Regeneration: Toronto's Waterfront and the Sustainable City: Final Report*. Toronto: RCFTW, 1992.

Ryder, Richard A. "Fisheries Inventories in Ontario: The Exigencies, Procedures and Ultimate Goals." April 5, 1960, in RG 1-282-1-24, "Lake Survey Procedures," 1961–1969, Ontario Provincial Archives.

———. "The Morphoedaphic Index—Use, Abuse, and Fundamental Concepts." *Transactions of the American Fisheries Society* 111 (1982): 154–164.

Ryder, Richard A., S. R. Kerr, Kenneth H. Loftus, and Henry A. Regier. "The Morphoedaphic Index, a Fish Yield Estimator—Review and Evaluation." *Journal of the Fisheries Research Board of Canada* 31 (1974): 663–688.

Sabatier, Paul A., and Hank C. Jenkins-Smith. *Policy Change and Learning: An Advocacy Coalition Approach*. Boulder, Colo.: Westview, 1993.

Salisbury, Edward. "The Study of British Vegetation." *Nature* 144 (1939): 305–306.

———. "The Origin and Early Years of the British Ecological Society." *Journal of Ecology* 52, suppl. (1964): 13–18.

Schiff, Ashley L. *Fire and Water: Scientific Heresy in the Forest Service*. Cambridge: Harvard University Press, 1962.

Schumacher, E. F. *Small is Beautiful: A Study of Economics as if People Mattered.* London: Sphere, 1974.

Sears, Paul B. "Ecology—A Subversive Subject." *BioScience* 14, no. 7 (1964): 11–13.

Seidel, Robert W. "A Home for Big Science: The Atomic Energy Commission's Laboratory System." *Historical Studies in the Physical Sciences* 16 (1986): 135–175.

Select Committee on Estimates. *Seventh Report, Session 1957–58; Nature Conservancy.* London: HMSO, 1958.

Shabecoff, Philip. *A Fierce Green Fire: The American Environmental Movement.* New York: Hill and Wang, 1993.

Sheail, John. *Nature in Trust: The History of Nature Conservation in Britain.* Glasgow: Blackie, 1976.

———. *Rural Conservation in Inter-War Britain.* Oxford: Clarendon, 1981.

———. "Nature Reserves, National Parks, and Post-war Reconstruction, in Britain." *Environmental Conservation* 11 (1984): 29–34.

———. *Pesticides and Nature Conservation: The British Experience 1950–1975.* Oxford: Clarendon, 1985.

———. "Grassland Management and the Early Development of British Ecology." *British Journal for the History of Science* 19 (1986): 283–299.

———. *Seventy-Five Years of Ecology: The British Ecological Society.* Oxford: Blackwell, 1987.

———. "Applied Ecology and the Search for Institutional Support." *Environmental Review* 13, no. 2 (1989): 65–79.

Shoard, Marion. *The Theft of the Countryside.* London: Temple Smith, 1981.

Shrader-Frechette, Kristin S., and E. D. McCoy. *Method in Ecology: Strategies for Conservation.* Cambridge: Cambridge University Press, 1993.

Shrubsole, Dan. "The Evolution of Public Water Management Agencies in Ontario: 1946 to 1988." *Canadian Water Resources Association Journal* 15 (1988): 49–66.

Shugart, Hank H., Robert A. Goldstein, Robert V. O'Neill, and J. B. Mankin. "TEEM: A Terrestrial Ecosystem Energy Model for Forests." *Oecologia Plantarum* 9 (1974): 231–264.

Siccama, Thomas G., F. Herbert Bormann, and Gene E. Likens. "The Hubbard Brook Ecosystem Study: Productivity, Nutrients and Phytosociology of the Herbaceous Layer." *Ecological Monographs* 40 (1970): 389–402.

Slobodkin, Lawrence B. "Intellectual Problems of Applied Ecology." *BioScience* 38 (1988): 337–342.

Smith, Roger, and Brian Wynne, eds. *Expert Evidence: Interpreting Science in the Law.* London: Routledge, 1989.

Smith, Stanford H. "Species Succession and Fishery Exploitation in the Great Lakes." *Journal of the Fisheries Research Board of Canada* 25 (1968): 667–693.

Smith, William H., F. Herbert Bormann, and Gene E. Likens. "Response of Chemoautotrophic Nitrifiers to Forest Cutting." *Soil Science* 106 (1968): 471–473.

Smolowe, Jill. "Con Ed to Drop Storm King Plant as Part of Pact to Protect Hudson." *New York Times,* 20 December 1980.

Society for Freedom in Science. *The Society for Freedom in Science, Its Origin, Objects and Constitution*. Society for Freedom in Science, 1953.

Sollins, Philip. "Organic Matter Budget and Model for a Southern Appalachian Liriodendron Forest." EDFB Memo Report 71–86, 1972.

Stannard, J. Newell. *Radioactivity and Health: A History*. DOE/RL/01830–159, October 1988.

Stapledon, R. George. "Cocksfoot Grass (*Dactylis glomerata* L.): Ecotypes in Relation to the Biotic Factor." *Journal of Ecology* 16 (1928): 71–104.

Steen, Harold K. *The U.S. Forest Service: A History*. Seattle: University of Washington Press, 1976.

Steidtman, James R. "The Geochemical Cycle of Iodine in Hubbard Brook Watershed." M.S. thesis, Dartmouth College, 1962.

Steinhart, John S., and Stacie Cherniack. *See* Executive Office of the President, Office of Science and Technology.

Stine, Jeffrey R. *A History of Science Policy in the United States, 1940–1985*. Washington, D.C.: Government Printing Office, 1986.

Sturges, Frank W., Richard T. Holmes, and Gene E. Likens. "The Role of Birds in Nutrient Cycling in a Northern Hardwoods Ecosystem." *Ecology* 55 (1974): 149–155.

Suter, Glenn W. "Ecosystem Theory and NEPA Assessment." *Bulletin of the Ecological Society of America* 62 (1981): 186–192.

Talbot, Allan R. *Power Along the Hudson: The Storm King Case and the Birth of Environmentalism*. New York: Dutton, 1972.

Tamplin, Arthur R. "Issues in the Radiation Controversy." *Bulletin of the Atomic Scientists* 27 (1971): 25–27.

Tansley, Arthur G. "Presidential Address." *Journal of Ecology* 2 (1914): 194–202.

———. *The Future Development and Functions of the Oxford Department of Botany*. Oxford: Clarendon, 1927.

———. "The Use and Abuse of Vegetational Concepts and Terms." *Ecology* 16 (1935): 284–307.

———. "British Ecology During the Past Quarter-Century: The Plant Community and the Ecosystem." *Journal of Ecology* 27 (1939): 513–530.

———. *The British Islands and Their Vegetation*. Cambridge: Cambridge University Press, 1939; reprinted 1953.

———. "Natural and Semi-Natural British Woodlands." *Forestry* 14 (1940): 1–21.

———. "The Values of Science to Humanity." *Nature* 150 (1942): 104–110.

———. *Our Heritage of Wild Nature: A Plea for Organized Nature Conservation*. Cambridge: Cambridge University Press, 1945.

———. "[Discussion of Fundamental Research in Relation to the Community]." *Advancement of Science* 3 (1946): 294–295.

———. "The Early History of Modern Plant Ecology in Britain." *Journal of Ecology* 35 (1947): 130–137.

———. "What Is Ecology?" *Biological Journal of the Linnean Society* 32 (1987): 5–16. Originally published in 1951 as a pamphlet by the Council for the Promotion of Field Studies.

Tansley, Arthur G., and T. F. Chipp, eds. *Aims and Methods in the Study of Vegetation*. London: British Empire Vegetation Committee, 1926.

Taylor, L. Roy. "Objective and Experiment in Long-Term Research." In *Long-Term Studies in Ecology: Approaches and Alternatives,* ed. Gene E. Likens, 20–70. New York: Springer-Verlag, 1989.

Taylor, Peter J. "Technocratic Optimism, H. T. Odum, and the Partial Transformation of Ecological Metaphor After World War II." *Journal of the History of Biology* 21 (1988): 213–244.

Teal, John M. "Energy Flow in the Salt Marsh Ecosystem of Georgia." *Ecology* 43 (1962): 614–624.

Thomas, Keith. *Man and the Natural World*. New York: Pantheon, 1983.

Thorpe, F. J. "Historical Perspective on the 'Resources for Tomorrow' Conference" (Resources for Tomorrow: Conference Background papers, 1961), 1: 1–13.

Tobey, Ronald C. *Saving the Prairies: The Life Cycle of the Founding School of American Plant Ecology, 1895–1955*. Berkeley: University of California Press, 1981.

"Tree Cutting and Nutrient Loss." *Science News* 99 (1971): 272.

Udall, Stewart L. "A Message for Biologists." *BioScience* 14, no. 11 (1964): 17–18.

United Kingdom, Advisory Council on Scientific Policy. *Annual Report, 1962–63*. Cmnd. 2163. 1963.

United Kingdom, Committee on Land Utilisation in Rural Areas. *Report*. London: HMSO, 1942.

United Kingdom, Forestry Commission. *Post-War Forest Policy*. London, 1943.

United Kingdom, Nature Conservancy. *Annual Reports of the Nature Conservancy*. London: HMSO, 1949–1960.

United Kingdom, Wild Life Conservation Special Committee. *Conservation of Nature in England and Wales*. London: HMSO, 1947.

United States, Atomic Energy Commission. *Annual Reports*.

———. "Jurisdiction of the AEC and Other Federal and State Agencies with Respect to Nuclear Power Facilities." Washington, D.C., n.d.

———. *Nuclear Power and the Environment*. Washington, D.C., 1969.

U.S. Congress, House Committee on Science and Astronautics, Subcommittee on Science, Research, and Development. *Environmental Pollution: A Challenge to Science and Technology,* 89th Congress, 2nd session, 1966.

———. *The International Biological Program: Its Meaning and Needs*. Washington, D. C.: Government Printing Office, 1968.

U.S. Congress, Senate Subcommittee on Public Lands. *"Clear-Cutting": Practices on National Timberlands, Part 1*. 92nd Congress, 1st session, 1971.

Van Dyne, George M. "Application of Some Operations Research Techniques to Food Chain Analysis Problems." *Health Physics* 11 (1965): 1511–1519.

———. "Ecosystems, Systems Ecology, and Systems Ecologists." Oak Ridge National Laboratory, ORNL-3957, 1966.

———. "Systems Ecology: The State of the Art." In *Environmental Sciences Laboratory Dedication, February 26 and 27, 1979,* ed. Stanley I. Auerbach and N. T. Millemann, 81–104. Oak Ridge National Laboratory, ORNL-5700, 1980.

Van Winkle, Webster. "The Application of Computers in an Assessment of the Environmental Impact of Power Plants on an Aquatic Ecosystem." In *Proceedings, ERDA-Wide Conference on Computer Support of Environmental Sciences and Analysis*, 85–108. Livermore, Calif.: Lawrence Livermore Laboratory, 1976.

Van Winkle, Webster, Lawrence W. Barnthouse, B. L. Kirk, and D. S. Vaughan. "Evaluation of Impingement Losses of White Perch at the Indian Point Nuclear Station and Other Hudson River Power Plants." Oak Ridge National Laboratory, ORNL/NUREG/TM-361 and NUREG/CR-1100, 1980.

Van Winkle, Webster, Sigurd W. Christensen, and Jack S. Mattice. "Two Roles of Ecologists in Defining and Determining the Acceptability of Environmental Impacts." *International Journal of Environmental Studies* 9 (1976): 247–254.

Varley, G. C. "Ecology as an Experimental Science." *Journal of Ecology* 45 (1957): 639–648.

Vig, Norman J. "Policies for Science and Technology in Great Britain: Postwar Development and Reassessment." In *Science Policies of Industrial Nations: Case Studies of the United States, Soviet Union, United Kingdom, France, Japan, and Sweden*, ed. Dixon Long and Christopher Wright, 59–109. New York: Praeger, 1975.

Vitousek, P. M., and W. A. Reiners. "Ecosystem Succession and Nutrient Retention: A Hypothesis." *BioScience* 25 (1975): 376–381.

Vogel, David. *National Styles of Regulation: Environmental Policy in Great Britain and the United States*. Ithaca: Cornell University Press, 1986.

Walker, J. Samuel. "Nuclear Power and the Environment: The Atomic Energy Commission and Thermal Pollution, 1965–1971." *Technology and Culture* 30 (1989): 964–992.

———. "The Atomic Energy Commission and the Politics of Radiation Protection, 1967–1971." *Isis* 85 (1994): 57–78.

Walsh, John. "Federal Laboratories: Report Asks More Interagency Research." *Science* 162 (1968): 442–444.

Watt, Alex. "Pattern and Process in the Plant Community." *Journal of Ecology* 35 (1947): 1–22.

Watt, Kenneth E. F. "The Choice and Solution of Mathematical Models for Predicting and Maximizing the Yield of a Fishery." *Journal of the Fisheries Research Board of Canada* 13 (1956): 613–645.

———. "Studies on Population Productivity II. Factors Governing Productivity in a Population of Smallmouth Bass." *Ecological Monographs* 29, no. 4 (1959): 367–392.

Webster, Andrew. *Science, Technology, and Society*. New Brunswick, N.J.: Rutgers University Press, 1991.

Weinberg, Alvin M. "Impact of Large-Scale Science on the United States." *Science* 134 (1961): 161–164.

———. "Nuclear Energy and the Environment." *Bulletin of the Atomic Scientists*, June 1970, 69–74.

———. *Reflections on Big Science*. Cambridge: MIT Press, 1971.

———. "Social Institutions and Nuclear Energy." *Science* 177 (1972): 27–34.

Weiss, C. "Research for Policy's Sake: The Enlightenment Function of Social Research." *Policy Analysis* 3 (1977): 531–545.

Werskey, Gary. *The Visible College*. London: A. Lane, 1978.

Weston, Sandra M. C. *The Canadian Federal Environmental Assessment and Review Process: An Analysis of the Initial Assessment Phase*. Ottawa: Canadian Environmental Assessment Research Council, 1991.

Whicker, F. Ward, and Vincent Schultz. *Radioecology: Nuclear Energy and the Environment*. Boca Raton, Fla.: CRC, 1982.

Whillans, Thomas H., Henry A. Regier, W. Jack Christie. "F. E. J. Fry's Field Studies: Good Field Data Provoke New Questions." *Transactions of the American Fisheries Society* 119 (1990): 574–584.

White, Gilbert F. "The Role of Scientific Information in Anticipation and Prevention of Environmental Disputes." In *Geography, Resources, and Environment: Selected Writings of Gilbert F. White*, ed. Robert W. Kates and Ian Burton, 377–392. Chicago: University of Chicago Press, 1986. Originally presented at the Conference on Adjustment and Avoidance of Environmental Disputes, Bellagio, Italy, 19–23 July 1974.

———. "Environment." *Science* 209 (1980): 183–190.

Whitaker, Herschel. "Early History of the Fisheries of the Great Lakes." *Transactions of the American Fisheries Society* 21 (1892): 163–179.

Whittaker, Robert H., F. Herbert Bormann, Gene E. Likens, and Thomas G. Siccama. "The Hubbard Brook Ecosystem Study: Forest Biomass and Production." *Ecological Monographs* 44 (1974): 233–254.

Woodwell, George M. "A Confusion of Paradigms (Musings of a President-Elect)." *Bulletin of the Ecological Society of America* 57, no. 4 (1976): 8–10.

———. "BRAVO Plus 25 Years." In *Environmental Sciences Laboratory Dedication, February 26 and 27, 1979*, ed. Stanley I. Auerbach and N. T. Millemann, 61–64. Oak Ridge National Laboratory, ORNL-5700, 1980.

Woodwell, George M., and H. H. Smith, eds. *Diversity and Stability in Ecological Systems: Report of Symposium Held May 26–28, 1969*. Brookhaven National Laboratory, BNL 50175 (C-56), 1969.

Worster, Donald. *Nature's Economy: A History of Ecological Ideas*. Cambridge: Cambridge University Press, 1977, reprinted 1985.

———. "The Ecology of Order and Chaos." *Environmental History Review* 14, nos. 1–2 (1990): 1–18.

Worthington, E. Barton. *The Evolution of IBP*. Cambridge: Cambridge University Press, 1975.

———. *The Ecological Century: A Personal Appraisal*. Oxford: Oxford University Press, 1983.

Yapp, William. "Introduction." In *The Biological Productivity of Britain*, ed. William Yapp and W. Brunsdon, ix–xii. London: Institute of Biology, 1958.

Index

Acid rain, 136, 168
Agricultural Research Council, 33, 34, 35, 36, 56, 57, 190
Agriculture: views of, in Britain, 60
Agriculture, Ministry of, 41
Aims and Methods in the Study of Vegetation, 15–16
Algonquin Park, 154, 156, 161, 171, 229
Allee, Warder Clyde, 180
Argonne National Laboratory, 67, 91
Atomic Energy Act, 66, 81
Atomic Energy Commission (AEC): and ecological research, 4, 64, 66–67, 68, 83, 87, 91–92, 118, 184, 198, 204, 216; political context of, 65; and nuclear technology, 65, 66, 83, 88; and environmental concerns, 82, 86, 87, 89–91, 104, 188, 215; and environmental impact assessment, 101–102, 106–108
Atomic Safety and Licensing Board, 102, 107, 111
Auerbach, Stanley: and ecology at Oak Ridge, 4, 70, 71–75 passim, 81–85, 103–104, 113, 190, 216; career, 66–68; and health physicists, 68–69; field studies, 69, 125; and ecosystem approach, 71; and radioecology, 79, 83; and International Biological Program, 93–98

Baker, John, 26, 27, 28
Ball, Sydney, 82
Balogh, Brian, 91
Beamish, Richard, 168
Bensley, B. A., 153
Biological Board of Canada, 153, 154
Birge, Edward, 120, 125
Blackman, F. F., 159

Bormann, F. H.: at Hubbard Brook, 4, 121, 124–128, 129–130, 223–224; career, 117–118, 134; and ecosystem ecology, 122, 123, 129–130, 138, 139–143; on ecology and politics, 131–133, 139–144; on ecology and forestry, 144, 145, 146
Botkin, Daniel, 137
Bowers, J. B., 31
Bramwell, Anna, 232
Brinkhurst, Ralph, 168
British Ecological Society, 15, 16, 22, 23, 25, 26, 29, 34
British Islands and Their Vegetation, 20, 28
British Trust for Ornithology, 30
Brookhaven National Laboratory, 121, 137
Bureau of Animal Population, 16, 30, 35
Burgess, Robert, 96
Burke, O. W., 98

Caldwell, Lynton, 194
Calvert Cliffs power station, 101
Capstick, C. K., 47
Carson, Rachel, 1, 181, 193, 233–234
Chant, Donald, 168
Christensen, Sigurd, 107
Christie, Jack, 164, 166, 173, 230
Clarke, C. H. D., 164, 165, 166
Clemens, W. A., 153, 154
Clements, Frederic, 18, 74, 193
Climate change, 199
Coleman, E. A., 126
Commission of Conservation, 153
Committee on Land Use in Rural Areas (Scott Committee), 21, 25
Commoner, Barry, 1, 181
Conference on Nature Preservation in Post-War Reconstruction, 22

Conservation: in Britain, 23, 32–33, 41, 57, 155; in Ontario, 155
Conservation movement, 162, 180, 183
Conway, Verona, 44, 213
Cooper, Charles, 132
Cornell University, 134
Council for Environmental Quality, 110
Coventry, Alan F., 155
Covington, W. W., 142, 144
Coweeta Hydrologic Laboratory, 93, 117, 118, 125
Cragg, J. B., 51, 213
Cronon, William, 205
Crossley, D. A., 76, 79, 80
Curlin, James, 80–81, 85
Curry, Robert, 130

Dansereau, Pierre, 184
Darling, Frank Fraser, 1, 193
Dartmouth College, 116, 118, 121, 124, 134
DDT, 98, 154
Dillon, Peter, 168
Diver, Cyril, 22, 30, 40, 210
Dominski, Tony, 128, 142, 144
Dower, John, 30, 33
Duffield, Robert, 91
Dunaway, Paul, 77, 79
Dunham, Charles, 217
Dunlap, Thomas, 181
Dymond, John, 155, 161, 162

Ecological Society of America, 185
Ecologists in Britain, 14, 15, 19–32 passim, 35, 37, 47–48, 58, 59
Ecology: and environmental politics, ix, 2, 6–8, 184–186, 190, 191, 196–205; social role, 2, 3, 123, 180–182, 192; growth of, 8, 183–185; and other disciplines, 189–191; in Canada, 232
Ecosystem concept: history of, 18, 71–74; relation to British ecology, 19, 51; theory, 84, 94–95, 98–99,

129–130, 172, 176; role in environmental politics, 109, 112–113, 114, 192, 199, 204; distinctive to ecology, 191
Edward Grey Institute, 30
Ehrlich, Paul, 181
Elliott, R. J., 52
Elton, Charles, 16, 18–20, 29, 30, 33, 35, 159, 210
Energy, Department of, 92, 110, 194
Environment, Ministry of, 168
Environment Canada, 173, 194
Environmental politics, 7, 41–42, 97, 103–104, 110–112, 114, 187, 191–201, 232
Environmental Protection Agency, 108, 110, 111, 145, 194, 197

Federer, Anthony, 126, 129, 142
Fish culture, 158–159, 162
Fisher, Stuart, 138
Fisheries management, 152, 156, 172–175, 230
Fisheries Research Board of Canada, 169, 170, 193
Forbes, Stephen, 153
Forcier, Larry, 136
Forest Service, 4, 119, 125, 126, 128–134, 146
Forestry Commission, 17, 18, 21, 41, 54, 56
Forman, Paul, 87
Freshwater Biological Association, 42
Fry, Fred, 154, 156, 158, 159–160, 162, 163, 167, 229
Fryer, John, 34
Fundamentals of Ecology, 49, 71, 73, 117–118, 120, 192

Game and Fisheries, Department of, 155–156
Georgia, University of, 93
Gilbert, O. J. W., 47

Gilmour, John S. L., 31–32
Gisburn Forest, 47, 48, 57
Godwin, Harry, 35
Gofman, John, 89
Goldschmidt, Victor, 72
Golley, Frank, 51
Goodwin, Harry, 23
Goodyear, Philip, 107
Gosz, James, 128, 142
Graham, Edward, 23
Great Lakes, 5, 151–178 passim

Hagen, Joel, 18, 78, 142, 182
Hall, Charles, 107
Harkness, William, 153–154, 156, 158, 164
Harvey, Harold, 168
Harwood, Jonathan, 190–191
Hasler, Arthur, 120, 125, 233
Hays, Samuel, 7
Holifield, Chet, 91
Holmes, Richard, 138, 139
Hornbeck, James, 126, 129, 133, 142
Hubbard Brook Experimental Forest,
    118–119
Hubbard Brook study: ecosystem
    ecology at, 5, 121–122, 139–142;
    relation of ecology to environmental
    issues at, 6, 117, 133, 142–144,
    199–201; compared with Oak Ridge,
    81, 100, 117, 123, 124, 125, 131–132,
    137, 139, 144–145; size of, 116, 142;
    origins, 117–124, 223–224;
    compared with Nature Conservancy,
    117, 123–125, 139; watershed
    experiments, 122–130, 134–135, 188,
    233; relation of ecology to forestry at,
    126, 130–136, 144, 146, 188–189,
    235; organization of, 134, 139;
    ecological models at, 136–137, 141,
    142, 227–228; ecology at, 138; as big
    science, 142, 227; compared with
    University of Toronto/Ontario
    government, 152, 153, 157, 170, 202

Hudson River, 4, 106–109, 111, 185,
    196, 198, 200
Huntsman, A. G., 164
Hutchinson, G. Evelyn, 49, 72–73, 74
Huxley, Julian, 23, 30, 32, 63

Institute for Terrestrial Ecology, 60, 234
International Biological Program,
    50–51, 80, 81, 88, 112, 114, 139, 142,
    169, 170, 180, 185, 193, 220, 230–231

Janak, James, 137
Johnson, Lyndon, 193, 195
Johnson, Noye, 116, 121–124, 127
Johnston, R. N., 156, 163
Joint Committee on Atomic Energy, 65,
    86, 90–92
Juday, Chancey, 120, 125

Kaye, Stephen, 77, 79, 82, 105
Kennedy, John F., 193
Krumholz, Louis, 66, 69
Kwa, Chunglin, 84, 97, 180, 182

Lake Erie, 154, 156, 158, 164, 165, 166,
    170, 171, 176
Lake Huron, 156, 158, 162, 164, 170,
    176
Lake Ontario, 154, 156, 158, 164, 166,
    170, 176
Lake Superior, 157, 176
Lands and Forests, Department of, 9,
    156, 161–166, 177
Langford, R. R., 161, 167
Larkin, Peter, 167
Laverack, M. S., 47
Leopold, Aldo, 162
Levins, Richard, 185–186
Likens, Gene: early work at Hubbard
    Brook, 4, 121; career, 119–121, 134;
    and ecosystem theory, 122, 129–130,
    138, 139–143; view of ecology, 123;
    and forestry, 131, 144–146; and

environmental politics, 131–133, 142–144, 145; on independence of researchers, 139

Lilienfeld, Robert, 97

Lindeman, Raymond, 71–72

Loftus, Ken, 173, 176

Loucks, Orie, 98

Lovins, Amory, 198

Lowe, Philip, 59

McElroy, William, 133

McEvoy, Arthur, 177

McGucken, William, 29

McIntosh, Robert, 179

Maple research station, 156, 159–160, 171

Margalef, Ramon, 73, 172

Marks, Peter, 135, 139, 142

Marsh, George Perkins, 119

Martin, Wayne, 129, 133

Marx, Leo, 180

Mazuzan, George, 87

Merlewood research station, 44, 46–47, 50, 52, 54, 56–59, 84–85

Mitman, Gregg, 180

Models, ecological. *See* Hubbard Brook study; Oak Ridge National Laboratory

Monks Wood Experimental Station, 50

Moore, Norman, 59, 215

Morgan, Karl, 66, 68

Morrison, Herbert, 32, 36, 40

Mountains and Moorlands, 44, 53–54

Muir, John, 7, 183

Naess, Arne, 232

National Environmental Policy Act, 88, 93, 101, 104, 114

National Parks and Access to the Countryside Bill, 36

National Parks Commission, 33, 39

National Research Council (Canada), 157, 162

National Science Foundation: support for Hubbard Brook study, 4, 117, 118, 121, 125, 132–133, 146, 188–189, 224; support for environmental research, 93, 132, 145, 184

National security: and ecology, 65, 87–88, 233

National Trust, 21, 22

Natural Environment Research Council, 60, 194

Natural Resources, Ministry of, 173, 176, 177

Nature Conservancy: origin and growth, 3, 14, 36–37, 39, 43, 211; significance to ecology, 5, 183–184, 190, 191; combining research and management, 6, 14, 52, 60; role of ecologists in, 14, 38, 60; and conservation, 39, 42–43, 57–58; funding of, 39, 40, 211; research of, 40, 41, 43, 57, 58, 214–215; autonomy of, 40–41, 56, 58, 59, 211, 212, 214–215; role in environmental politics, 41–42, 53, 56, 58, 186, 234; Sites of Special Scientific Interest, 42; nature reserves, 52–53, 186, 209; and other agencies, 56–57; compared with Oak Ridge, 64, 65, 68, 69–70, 75, 84–86, 87, 97–98, 103; compared with Hubbard Brook, 117, 123, 124–125, 139; compared with University of Toronto/Ontario government, 154, 155, 162

Nature Conservancy Council, 60

Nature reserves: Kingley Vale, 13; Wicken Fen, 23; Woodwalten Fen, 23; Beinn Eighe, 43; Yarner Wood, 43; Moor House, 43, 44, 49, 57; Roudsea Wood, 47, 51–52, 58, 213; Holme Fen, 48

Nature Reserves Investigation Committee, 22, 30, 31, 42

Nelson, Daniel, 70, 80–81, 85

Nicholson, Max: and ornithology, 30, 32; career, 30, 40; and Wild Life Conservation Special Committee, 30–31; and Nature Conservancy, 36, 40, 41; and conservation, 41, 42, 53, 59, 187; on ecology, 47; and Political and Economic Planning (PEP), 55, 214; and basic research, 57; view of nature, 232

Nixon, Richard, 195

Nuclear Regulatory Commission, 92, 102, 110

Oak Ridge National Laboratory: growth of ecology at, 4, 64, 75–76, 110, 184, 190–191; IBP at, 4, 93–101, 196; significance to ecology, 5; relation of ecology to environmental issues at, 6, 79, 80, 98, 99, 101–103, 110, 111, 114, 187–188, 198–199, 222; origins of, 9, 63; and science policy, 64–65; compared with Nature Conservancy, 64, 65, 68–70, 75, 84–87, 97–98, 103; autonomy of research at, 65, 85–86; radiation risks at, 65–66; and waste disposal, 66, 86–87; health physics at, 66, 85, 86, 190; relations between ecology and health physics, 67–68, 70, 190; relations between ecology and physical sciences at, 69, 82, 190; and ecosystem concept, 71, 82, 84, 114, 190; place of ecology in, 74–76; ecological research at, 76–85; radioecology at, 77–78, 79, 83; ecological models at, 78–79, 95, 104–105, 107, 109, 182, 220; Walker Branch study at, 80–81, 83, 187; compared with Hubbard Brook, 81, 117, 123, 124, 125, 131–132, 137, 139, 144–145; contribution of ecology to AEC missions at, 81, 82, 83, 88, 188; and Eastern Deciduous Forest Biome, 93–94, 95–96, 100,

102, 113, 221; ecology as big science at, 96; new ecology building at, 113; and Environmental Sciences Division, 114; compared with University of Toronto/Ontario government, 152–153, 169, 170; environmental mandates of, 196; mentioned, 191, 200, 205

Odum, Eugene, 70–71, 117–118, 120

Odum, Howard T., 73, 180, 181–182

Olson, Jerry, 77, 78–79, 80, 83–84, 85, 93

O'Neill, Robert, 96, 98, 220

Ontario Fisheries Research Laboratory, 153, 154, 156, 158, 228

Ontario Hydro, 170

Oosting, Heinie, 117

O'Riordan, Timothy, 7, 232

Ovington, J. Derek, 39, 44–51, 53–55, 57–59, 69–70, 84–85, 122, 213

Oxford Bird Census, 30

Park, Orlando, 66

Patten, Bernard, 73, 79, 80, 85

Pearsall, William, 20, 22, 45–46, 48–49, 57, 58, 118

Petulla, Joseph, 180–181

Pierce, Robert, 118, 119, 125, 126, 128, 130, 133–134, 142, 146

Pimlott, Douglas, 167

Polanyi, Michael, 27, 28

Pollution Probe, 168

Public Health Service, 86, 121

Regier, Henry, 164, 168, 171–172, 173, 176, 201

Reichle, David, 76, 79, 93, 96, 98, 99

Rigler, Frank, 167–168, 169

Riley, George, 49

Robinson, Roy, 17

Rothamstead Experimental Station, 16

Rothschild, Charles, 21

Ryder, Richard, 163

Salisbury, Edward, 20, 22
Salmonid Communities in Oligotrophic Lakes Symposium (SCOL), 171–173, 176
Satchell, J. E., 47
Schumacher, E. F., 198
Science policy, 8–9, 65, 87, 192, 195–196
Sears, Paul, 2
Sheail, John, 21, 34, 43, 59
Shelford, Victor, 181–182
Siccama, Tom, 137, 226
Smith, Herbert, 22, 31
Smith, Stan, 173
Smith, William, 127
Society for Freedom in Science, 27, 29, 36
Society for the Promotion of Nature Reserves (SPNR), 21–22
Soil Conservation Service, 119
South Bay research station, 156, 157–158, 161, 162
Stapledon, R. George, 16
Steidtman, James, 119, 224
Strategic Plan for Ontario Fisheries, 153, 175–177, 201
Struxness, Edward, 66, 68, 69
Sturges, Frank, 138
Systems analysis, 95, 97–99, 104, 194, 196

Tamplin, Arthur, 89
Tansley, Arthur: advocacy for Nature Conservancy, 3, 13, 30, 31, 37; on ecology and practical concerns, 15–16, 20–21, 23, 35, 55, 186; and Forestry Commission, 17; and ecosystem concept, 18, 19, 29, 45, 47, 51, 74; and conservation, 22–28 passim; and autonomy of science, 27, 28, 29, 35
Taylor, Peter, 180, 182
Teal, John, 73
Thetford Chase, 48
Thoreau, Henry David, 7, 183
Toronto, University of/Ontario government: ecological research, 5, 167–168, 169, 170, 191; fisheries management, 5, 6, 163–164; university/province relations, 5, 152, 154, 156–157, 167, 170–171; compared with Hubbard Brook, 152, 153, 157, 170, 161, 202; compared with Oak Ridge, 152–153, 161, 169, 170; fisheries research at, 153–154, 182; compared with Nature Conservancy, 154, 155, 161, 162; relation of ecology to fisheries management, 157, 162–163, 184, 189, 201, 229; autonomy of ecological research at, 162, 163; relation of ecology to environmental issues at, 168–170; ecologists at, 204, 205
Town and Country Planning, Ministry of, 29, 30, 31, 32, 34
Town and Country Planning Bill, 33

Udall, Stewart, 193

Van Dyne, George, 79–80
Varley, George, 48
Vernadsky, Vladimir, 72

Waid, J. S., 47
Walker, J. Samuel, 87
Wallis, James, 137
Watt, Alex, 18, 20, 140, 208, 227
Weinberg, Alvin, 69, 81, 83, 84, 91–92, 96–97, 187, 218–219
White, Gilbert, 7
Whittaker, Robert, 136, 137
Wild Life Conservation Special Committee, 23–24, 30–35, 42, 44
Wisconsin, University of, 93, 120
Wolfe, John, 66–67, 69, 70–71, 198
Woodwell, George, 185
Worster, Donald, 180, 181, 182
Worthington, E. B., 42, 43, 52, 57, 59

Yale University, 134

CPSIA information can be obtained
at www.ICGtesting.com
Printed in the USA
FSHW010355250119
55121FS